T0074856

ADVANCE PRAISE FOR *TOTAL GARBAGE*

"This brilliant book will first make you angry and then motivate you to do more to protect this blue planet on which we all depend. It exposes the bafflingly wasteful practices that have become commonplace in modern society and goes further to provide clear ways we can act right now to solve our most pressing environmental problems. From passive houses to induction stoves, refillable bottles, electric golf carts, and new laws that would force firms away from disposable, single-use packaging, we see here a mix of individual and collective actions that can make a real difference in the real world. The gas industry and businesses hoping to preserve the status quo will probably call this book total garbage, but don't buy what they're selling. The stories told here will open your eyes to the possibilities of a future in which kids don't get asthma from indoor gas ranges and where we avoid eating a credit card's worth of microplastics every week. There is hope in these pages, which is something those of us living on this warming planet need now more than ever."

—Bart Elmore, professor of environmental history at Ohio State
University and author of *Citizen Coke: The Making of Coca-Cola
Capitalism* and *Seed Money: Monsanto's Past and Our Food Future*

"Our system produces waste almost as if that was its intent, but there are always people thinking deeper and more clearly, and their stories will help you see the possibilities for a very different world!"

—Bill McKibben, author of *The End of Nature*

"Covering everything from your kitchen table to the depths of the ocean, this is the book to inform you about one of the biggest, most substantial crises humanity faces today: garbage. Thoughtful, inspiring, concerning, and even fun, Humes's book will entertain and inform you in ways that may well save both your life and our planet. Enjoy!"

—Thom Hartmann, radio personality and author of
The Last Hours of Ancient Sunlight

"Edward Humes's subject is waste, and he doesn't waste a word making his eye-opening case. He zeroes in on ordinary things—food, energy, buildings, cars, clothing—that have morphed into grotesque, life-threatening, planet-destroying problems. For instance, plastics—a brilliant, almost indestructible invention we're now ludicrously using to make things like potato chip bags that will be thrown away in minutes and last for centuries. But Humes is here to help, not horrify, and he introduces us to extraordinary, smart, indefatigable people who offer practical, affordable answers, concepts that in a few more years may well become household words—heat pumps, induction stoves, passive houses, community solar gardens, rural electric gardens, urban microfarms, refill stores, and thrifting. *Total Garbage* is total gold!"

—Tony Hiss, award-winning author of *Rescuing the Planet:
Protecting Half the Land to Heal the Earth*

"Humes is back to make you want to read about waste again. *Total Garbage* is like a constellation of stories about people who shine brightly on their own, and yet together illuminate a path forward on one of the most pressing issues of our time, making waste obsolete."

—Dr. Jenna Jambeck, MacArthur genius grantee and coauthor of *Plastics*

"Oh what a mess we've made. We are all villains in the story of *Total Garbage*, but Edward Humes's exhaustive reporting reveals that some among us—a genius trashologist in Georgia, a rebel farmer in Los Angeles, a small-town mayor in a golf cart—have become waste-fighting heroes, too. Read this book, and you'll know how to join their fight. You'll never look at a piece of plastic the same way."

—McKenzie Funk, journalist and author of *Windfall:*
The Booming Business of Global Warming

"Waste has been normalized to the point that many of us hardly notice it, but Edward Humes argues that by confronting our waste, we can address our intertwined global crises. You'll be inspired by the ordinary people filling *Total Garbage*—young and not-so-young, rural and urban, red state and blue—using creativity and common sense to tackle waste in all its forms, and thus tackle the biggest challenges we face. I love this book!"

—Anne-Marie Bonneau, author of *The Zero-Waste Chef*

"Every home with microplastics in the water from the tap, in the milk in the fridge, in the fruit on the counter—that is to say, all of them—should also have this book in it."

—Hope Jahren, author of *Lab Girl* and *The Story of More*

"Did you know that two-thirds of all energy produced is wasted, or that a third of all food ends up in landfills? Staggering wastefulness is the root problem of our biggest global crises, Edward Humes argues. By reframing issues like climate change, plastic pollution, and energy as waste problems, he finds a slew of hopeful solutions. *Total Garbage* will make you rethink how you live, shop, cook, eat, and travel. It's a keeper. And now I am shopping for a heat pump and an induction stove!"

—Susan Freinkel, author of *Plastic: A Toxic Love Story*

"In *Total Garbage*, Edward Humes talks trash about wastefulness, describing it as the underlying cause of the apparently disparate environmental crises that afflict the world. His diagnosis not only enables us to understand our predicaments more accurately but also empowers us to address them more effectively. Engagingly written and convincingly argued."

—Glenn Branch, deputy director, National Center for Science Education

TOTAL
GARBAGE

TOTAL GARBAGE

How We Can Fix Our Waste and Heal Our World

—————

EDWARD HUMES

Avery
an imprint of Penguin Random House
New York

AVERY

an imprint of Penguin Random House LLC
penguinrandomhouse.com

Photograph on pages iv–v, xi, 1, 63, 131 from Shutterstock

Most Avery books are available at special quantity discounts for bulk purchase for sales promotions, premiums, fundraising, and educational needs. Special books or book excerpts also can be created to fit specific needs. For details, write SpecialMarkets @penguinrandomhouse.com.

Library of Congress Cataloging-in-Publication Data

Names: Humes, Edward, author.
Title: Total garbage: how we can fix our waste and heal our world / Edward Humes.
Description: [New York]: Avery, an imprint of Penguin Random House, [2024] |
  Includes index.
Identifiers: LCCN 2023043542 (print) | LCCN 2023043543 (ebook) |
  ISBN 9780593543368 (hardcover) | ISBN 9780593543382 (epub)
Subjects: LCSH: Refuse and refuse disposal. | Waste minimization.
Classification: LCC TD791 .H86 2024 (print) | LCC TD791 (ebook) |
  DDC 628.4/4—dc23/eng/20231227
LC record available at https://lccn.loc.gov/2023043542
LC ebook record available at https://lccn.loc.gov/2023043543

Printed in the United States of America
1st Printing

Book design by Silverglass Studio

This book is dedicated to the generation that gave us Nalleli Cobo, Sydney Bauer, Greta Thunberg, the Montana 16, and so many others. Thank you for embracing action as the antidote to despair.

It's easy to say "It's not my child, not my community,
not my world, not my problem."

Then there are those who see the need and respond.
I consider those people my heroes.

—FRED ROGERS

Those who say it can't be done are usually interrupted by others doing it.

—JAMES BALDWIN

I don't want you to be hopeful. I want you to panic. I want you to feel the fear I feel every day. And then I want you to act.

I want you to act as you would in a crisis. I want you to act as if the house is on fire. Because it is.

—GRETA THUNBERG, AT AGE SIXTEEN

Contents

TOTAL GARBAGE

Prologue
The Credit Card

You swallowed 285 pieces of plastic today.

You will do it again tomorrow, and the next day, and the next. You've been doing this for years, and you will likely continue doing it for the rest of your life, like it or not. (Spoiler alert: there's nothing about this to like.)

This involuntary seasoning of plastic waste particles is in the food you eat and the water you drink—tap and bottled water alike. It's in the air, it's in the dust on your bookshelves, it's in the soil. Your salt shaker has plastic mixed in with the grains. Even your favorite IPA or sour ale gifts you with a little bit of plastic with each sip.

The World Wildlife Fund and Australia's University of Newcastle were kind enough to give us all indigestion by testing drinking water from all over the planet, then combining their findings with the results of other scientific studies of plastic ingestion around the world. This seemed like a good way to estimate how record amounts of plastic production—and record amounts of plastic waste—might be affecting not only land, sea, air, and our fellow creatures, but also our food, drink, bodies, and health.

What they found: there is no escaping plastic waste. Every human on the planet is in this same plastic boat. If you accidentally gulp a mouthful of seawater when you dive into the warm turquoise waters of the Gulf of Mexico, you're swallowing plastic. If you melt ice scooped from untrodden Alaskan tundra to drink, you'll find plastic in that, too. And when you turn on your tap, grab ice cubes from the fridge, or boil a pot of pasta, it's in there, and then in you, too. That's the new inverse rule of trash: garbage out, garbage in . . . us.

The researchers call out shellfish, beer, and salt as major culprits in plastic ingestion, but all pale in comparison to water. The number of plastic bits in drinking water runs highest in the United States of all countries tested, which makes sense, given that no other country produces or consumes more plastic beverage bottles than America. Tests found plastic in 94.4 percent of US water samples, averaging about five plastic fibers per 500 milliliters (that's a smidge under 17 ounces). Europe had the lowest combined score, though it's not much of a bragging right—72.2 percent of samples had plastic mixed in, with an average of just under two fibers each.

Country	Average No. of Fibers per 500 ml Tap Water	Tap Water Samples with Plastic Fibers
United States	4.8	94.4%
Lebanon	4.5	98%
India	4	82.4%
Uganda	2.2	80.8%
Ecuador	2.2	79.2%
Indonesia	1.9	76.2%
Europe	1.9	72.2%

Source: WWF International

Granted, these are very tiny amounts of plastic—mere microplastics, barely or not at all visible, a single millimeter long or less. And the daily intake of 285 bits is just an average estimate, so some days it's more, others less. Yet over time, they add up. On average, we each consume somewhere between five grams of plastic a year, and five grams a week.

How much plastic is five grams? Think of it as pulling a credit card out of your wallet, chewing it, and swallowing it. All of it. Up to once a week, every week. Forever.

It's as if we've all been lab rats in a secret multigenerational experiment: What happens if we are so reckless in policy and practice with our most toxic garbage that we all begin eating it without realizing it? No one knows with absolute certainty the answer to that question, but the implications of making a new and unintended food group out of plastic are undeniably ominous. As food, plastic is about as nutritious as eating your raisin-bran box instead of the cereal inside—and a lot less benign. Plastic is a fossil fuel product, and almost every type has been shown to contain potentially toxic chemicals, even types that have been declared safe to hold our food and clothe our bodies. There are at least 2,400 known toxins that can be released by plastic, including one class of substances called PFAS, also known as "forever chemicals" because they basically never go away, outside you or in. Plastic toxins have been linked to infertility and sexual dysfunction in adults, impaired physical and intellectual development in children, high blood pressure, weakened immune systems, and a variety of cancers—which have been on the rise for the past thirty years, especially in adults under fifty. At the top of this list: cancers of the digestive system.

There's no scientific proof that snacking on plastic causes these cancers, though the risks seem pretty obvious: Plastic + Toxins + Gulp, Gulp, Gulp = Nothing Good. And that thirty-year cancer time frame matches up with the rise of global plastic waste and pollution—now 400 million tons a year and counting.

So next time you grab a takeout coffee in a disposable "paper" cup that is actually a plastic-lined cup, spoon in sugar with a plastic utensil or tear open a sugar packet (plastic-coated paper), pour in cream from a plastic pitcher or creamer capsule, then bite into a pastry wrapped in plastic and finally wash it down with a swig through a plastic sippy-cup lid, remember all the plastic involved in this one café visit. Not a shred of it will be recycled no matter what bin you toss it in. Then consider that Starbucks alone sells 4 million cups of coffee *a day*, with sleeves and lids and those green plastic stoppers—and that's just one slice of our disposable plastic world. Now, perhaps, you can begin to see where that credit card we could be eating weekly comes from.

But it won't stay at that level forever. If the current growth in disposable plastic production and pollution continues, we'll be eating two credit cards a week in no time.

Okay, I know that this plastic-eating thing is beyond gross. Also bewildering. Perhaps enraging, or scary, or frustrating. Some of you might wish you had skipped over these pages. But now that this is stuck in your brain, what happens the next time you're feeling thirsty and you see all those frosty, overpriced plastic bottles of soda, "enhanced" water, and sweet iced tea in the grab-and-go coolers fronting every supermarket checkout lane? Do you stand your ground and say "no more" to that weekly credit card? Or do you make the impulse buy, because, really, one bottle more or less— or chip bag or candy wrapper or fill-in-the-blank piece of our disposable economy—won't make a difference either way? Except, deep down we know that's not true. We make dozens of choices like this every day, and so does everyone else, which means our choices *do* matter—because they all add up. One choice is a blow against the rising tide of plastic pollution. The other sends the message: *Bring on those two credits cards a week!*

Of course, that raises a bigger question: Is it even possible to do something about this? After all, it's not like plastic pollution is the only scary, seemingly unsolvable global environmental crisis we're facing. There's climate disruption. The energy crisis. Fossil fuel pollution. Wildfires. The farm and food crisis. A broken recycling system overflowing with trash. Cities so hot that grabbing the wrong doorknob or falling to the sidewalk means third-degree burns, a trip to the ER, and painful skin grafts to repair the damage. The environmental doomscroll goes on and on. And, seriously, doesn't it get to be too much—these faceless, overwhelming, anxiety-inducing forces that make you want to shut it all out and retreat into binge-watching *Ted Lasso* for the third time? It's hard for any of us to imagine doing something meaningful about even one of these crises, much less all of them.

Now, this is the part where I'm supposed to say, *But wait, there's hope!*

I'm not going to say that.

We've spent the last fifty years hoping that if we just ignore it, it'll go away, or perhaps some magical tech wizardry will appear for a big *ta-da!* moment and fix everything, or the powers that be, whoever and wherever they are, will solve it by fiat. That hasn't happened. It will never happen. Instead, we've been suckered by one impossible snake oil cure after another. The Hydrogen Highway! Ethanol! Chemical recycling! Clean gas! Clean coal! Carbon capture! They're all scams, greenwashing, or failures. Now the time for hope and waiting and pretending there's nothing we ordinary mortals can do about any of this is long past.

We are officially in the let's-get-off-our-asses-and-get-it-done-now-or-else moment. And the good news is that it *is* getting done. Not by hope or magic or royal edict, but with real solutions that have been hiding in plain sight for years, pioneered by ordinary folks doing extraordinary things. In neighborhoods and communities and on campuses across the nation, they're cracking the code on climate, plastic pollution, energy—all the big crises—not as separate problems but as symptoms of one overarching disease:

Waste.

This is the true battle, challenge, and opportunity of our age. Waste is the engine driving those seemingly unrelated, unstoppable global crises. From our trash cans to our tailpipes, from our stoves to our plastic soda bottles, from our power plants to what we plant, we have unwittingly become the most wasteful civilization in history: wasteful of food, energy, resources, fuel, material—everything. Simple, prosaic, heedless waste, trashiness in its innumerable forms, is our archvillain.

Oddly enough, the fact that there is one root crisis to rule them all is a good thing. Waste is the one big economic, environmental, and social justice problem *anyone* can do something about—and see immediate payoffs in prosperity, health, and quality of life even while saving the planet. It's not just waste that adds up bit by bit, with our unintended plastic credit card diet being exhibit A. Fixing waste yields equal and opposite additive rewards as well.

This is not mere semantics. Reframing those paralyzing environmental crises as symptoms of the waste we can see, touch, and change around us is the secret sauce that turns helplessness and anxiety into opportunity. "Waste," after all, is synonymous with "cost," with "mess," with "inefficiency" and "loss"—things everyone can agree are bad. When waste is the common enemy, fixing the planet stops being about giving up things we love. Instead, it's about *upgrading* to stuff we'll love better. Because not wasting our stuff *is* better.

But . . . waste is an elusive foe. It's so big, so deeply embedded in our economy, our marketing, our most beloved brands, and our daily lives, that it's difficult to see clearly, or to see at all. Even so, growing numbers of families, neighborhoods, campuses, and communities around the world are doing just that. They are rethinking our "normal" ways, our customs and habits, our acceptance of inherently wasteful things because "that's the way we've always done it" and "that's just the way things are." And they're starting to see these "normal" things as kind of abnormal, and even a little bit insane.

They wonder how it can be that the average American now creates 8.2 pounds of garbage every day, 1.5 tons a year—three times our trashiness in 1960, and a pound more a day than just ten years ago.

They question why we love "cooking with gas" so much that it's a meme for when everything's going great—even though gas stoves are antiquated, wasteful energy hogs that spew enough indoor pollution to drive a 42 percent higher risk of childhood asthma in homes that have them, while constantly leaking methane, one of the most potent heat-trapping pollutants in the world.

How, they ask, is it normal, rather than a march-on-Washington outrage, that America wastes two-thirds of its trillion-dollar annual energy supply, even though much of that waste—the cost, the pollution, the devastating effect on climate, the crazy-high electric bills we get in summer (and winter and, well, all the time now)—is avoidable?

Why do we let our washing machines flood rivers and oceans with microplastic waste, even though a simple filter could cut that plastic pollution off at the source?

Why do we habitually plant water-wasting, unproductive, nature-disrupting, non-native grass lawns in front of our houses, even in areas plagued by drought, or food shortages, or the mass demise of bees, fireflies, and other pollinators?

How is it that we accept without a thought the disposable bottles, cups, bags, and packaging that were roundly rejected by past generations of American consumers? We had to be trained by decades of marketing to see "convenience" where once we saw waste, and to allow industries to foist the cost of the unrecyclable mess they created onto taxpayers. Now all that instant plastic trash is everywhere and in everything, polluting our water, our land, our food, and our bodies. Microplastics are even found by cardiac surgeons inside human hearts, and by after-birth testing of women's placentas, amniotic fluid, and the first poops of newborns. Is "insane" too strong a word for this brand of "normal"? Or not strong enough?

Then there is the single most wasteful product in our daily lives: our gasoline-powered cars. Sacrilege, I know. But seriously, they waste 80 percent of the fuel we buy—$500 billion up in smoke every year in fuel costs in America alone, a level of wastefulness that's almost unchanged since the days of the Ford Model T. That grade of 20 out of 100 in terms of efficiency is a feature of the physics of internal combustion (versus 90 percent efficiency for electric cars). Such performance would be an abject fail in any other realm, an F– grade we might well ground our kids for getting in school, one that should have propelled us to innovate our way to something better long ago. Yet we have meekly accepted this epic, damaging, polluting, cancerous, lung-disease-causing, climate-wrecking, and costly wastefulness in our cars for more than a century. Why?

And why does it make us feel so defensive or angry to read this sort of thing—not at the shortcomings of our cars, but at the suggestion that they are objectively terrible? Isn't it kind of crazy to defend and love inanimate objects that can't love us back—cars and gas stoves and lawns and plastic soda bottles—so much that acknowledging their faults and imagining alternatives seems difficult, painful, infuriating, or downright impossible?

Is it any wonder, then, that waste in all its forms is now the biggest and most expensive thing we make? Or that we accept levels of wastefulness in our daily lives unprecedented in human history as normal, seductive, inevitable, or invisible?

In the pages that follow, you will read about people who are rejecting the normalcy of waste and reaping all sorts of benefits in the process. You'll meet residents of a community where people rarely use their gas-powered cars because they have something better, cheaper, less wasteful, more convenient . . . and way more fun (and no, this is not about simply swapping gas tanks for batteries). Then there's the farm town in frosty Minnesota that has partnered with the local university to go all in on renewables, making wind and sun one of their most important cash crops. They even power the local liquor store with solar ("We chill your beer with the sun!"). You'll meet the restaurant chef in Pennsylvania who's championing a more efficient, healthy, less-wasteful way to cook, and the former financier in Los Angeles who wants to turn your water-wasting lawn into an urban microfarm. Then there's America's official "Trash Genius," who's helping communities throughout the country and the world figure out where plastic pollution is coming from—and how to stop it. And you'll also meet the single mom in Maine who took on multinational beverage and snack food giants to reinvent our broken recycling system, forcing producers of plastic pollution to clean up their mess instead of taxpayers footing the bill for the disposable economy. That's one person with a seemingly impossible vision who has reset national policies and entire industries—and she's just getting started.

The common thread in these stories illustrates the central premise of *Total Garbage*:

1. Waste is public enemy number one.
2. Wasting less is our greatest untapped economic, environmental, health, and quality-of-life opportunity.

3. Wastefulness is driven by choice, habit, and marketing, rather than by necessity, inevitability, and economic sense. We are neither hopeless nor helpless to act.
4. Once you start seeing the waste embedded in our daily lives, you won't be able to *stop* seeing it everywhere. And once you start enjoying the benefits of the less-wasteful choices out there, you won't want to stop.
5. There are five aspects of our daily lives where opportunities to be less trashy abound: what we throw away; how we heat, cool, cook, and power our places; how we get around; the stuff we buy to wear and use; and what we eat and drink.

Getting started on the way back from total garbage doesn't have to be complicated or expensive. It doesn't mean everyone has to rush out to buy a $50,000 electric car most of us can't afford. It's not about everyone returning to the Stone Age, either, or going all granola and Birkenstocks, or sacrificing comfort and living off the grid. The perverse silver lining of our addiction to wastefulness is that it is literally everywhere, which means myriad opportunities are all around us to embrace low- or no-cost solutions to waste.

Change comes in two ways. It can be driven from the top down, with the sorts of new laws championed in Maine and elsewhere aimed at ending the free ride that makers of wasteful products have long enjoyed—a silent subsidy of waste. Polluters should pay, not taxpayers, is the theme here. Responsibility is being shifted to those who created, marketed, and profited from these products, then assured us it's fine, don't worry, it can all be handled, even as plastic and tailpipe and smokestack pollution poisons us and our world. But such new laws take time and face powerful opposition, and sometimes even commonsense and righteous attempts to hold polluters accountable never make it across the finish line.

Fortunately, change is also driven from the ground up, home by home, neighborhood by neighborhood, and community by community. This sort of tidal change is not so easy to stop with lobbyists, lawyers, and campaign donors. The personal payoffs are often more immediate, and the collective difference they can make is powerful.

Marine biologist, conservationist, and author Ayana Elizabeth Johnson put it this way in her stirring 2023 commencement address at Middlebury College:

> It is naive to expect that governments and corporations will do the right thing, or that someone else will handle it. It is naive to think we can "solve" or "stop" climate change. It is also naive to give up, when every tenth of a degree of warming we prevent, every centimeter of sea level rise we avoid, every species we save, and every increasingly unnatural disaster we avert all matter so very much.

In the early days of the environmental movement and the quest to do the right thing, there were the "3 Rs"—Reduce, Reuse, Recycle. Later it became five: *Refuse*, Reduce, Reuse, Recycle, *Rot*. That was better—the natural rot of compost is critical for dealing with food waste, and the simple act of just saying no, of refusing the waste of the disposable economy, desperately needed to be on the list.

But it was not enough. Those five worthy imperatives were too limited, taking aim at only products and food. But now we know that the waste driving our big environmental crises goes far beyond things we buy at the store. Our habits, our use of energy, and how we choose to get around have enormous impact—and are enormously wasteful. So there are three more Rs we have to add to the old list if we truly want to counter waste in all its most damaging forms.

We need to *Repair*. We've forgotten how to mend. It's time to relearn.

We need to *Repower*. Supporting and joining the shift from the world-crushing waste of fossil fuels to cheap, clean, renewable energy is vital—using our voices, our votes and, whenever possible, our dollars, both public and private.

First and foremost, we need to *Rethink*. Wastefulness is not natural for humans, or any living creature. Yet it is, like any behavior, something we have learned, accepted, normalized, and turned into habit. Rethinking our waste and all its terrible costs—imagining, then embracing alternative choices and habits—is the first big step in kicking the total garbage crisis.

These are our new 8 Rs:

| Rethink | Repower | Refuse | Reduce |

| Reuse | Repair | Recycle | Rot |

So where do we begin? Maybe rethinking starts when your next-door neighbor asks you over for dinner and says, "Hey, check out our new gas-less induction stove. You won't believe how cool it is." Or when someone down the street shares some succulent tomatoes from their home garden, and you realize those red cannonballs from the supermarket you've been eating just can't compare. Or maybe it's when you see a family you know tooling around town in an electric golf cart, and when you ask about it, they laugh and tell you, sure, they still have their car, but they haven't been to the gas station in a month. Or just maybe your road to waste-mindfulness begins with a walk on the beach and the realization that all those colorful bits on the sand at the tide line aren't pretty shells, but fragments of plastic—and your kids don't know that's not normal. Or rather, it didn't used to be.

That's how change spreads and how we turn the tide on waste and its dire symptoms. When whole communities unite by word of mouth and common ground, these small changes begin to gather strength and add up to meaningful impact. Our choices matter: they can harm or heal the environment, they can drive those top-down government policies we need and overturn ones we don't, they can move markets, they can make wasteful and harmful products lose to beneficial and sustainable ones, and they can make, break, or remake economies. And they can help save the planet, too. That's how we take a stand against total garbage.

And it all starts with being less trashy.

PART I

Our Dirty Love Affair with Trash

We are now faced with the fact . . . that tomorrow is today. We are confronted with the fierce urgency of now. In this unfolding conundrum of life and history there is such a thing as being too late. Procrastination is still the thief of time. . . . Over the bleached bones and jumbled residues of numerous civilizations are written the pathetic words: "Too late." There is an invisible book of life that faithfully records our vigilance or our neglect. . . . Now let us begin. Now let us rededicate ourselves to the long and bitter—but beautiful—struggle for a new world.

—Dr. Martin Luther King Jr., Riverside Church, New York, April 4, 1967

1

Our Disposable Age

The innocent question that changed Ryan Metzger's life came the summer his son turned six. That's when Owen asked about the ever-expanding bag of old batteries in the junk drawer.

"What's going to happen to them, Dad?" he asked. "What are we supposed to do with them? We're learning about recycling in school. Where do *these* get recycled?"

"Um," Metzger said. "I don't know."

He knew where to *get* batteries, of course—everyone did. And there were always instructions on correctly inserting and using them. But instructions on what to do when they died? Not so much. That's why he fell into the habit of stuffing dead batteries into a drawer filled with all the other small, disused stuff that the family wasn't sure what to do with. As for bigger discards, there was a spot in the basement for those. And all the old paint cans and turpentine and lawn chemicals were collecting dust and rust out in the garage.

"It's heavy, Dad." Owen waved the bag of batteries around.

It was pretty full, Metzger had to admit. Detritus from flashlights and old toys, smoke alarms and remote controls, with a crusty one that came out of an old toothbrush, these batteries were one of many types of problematic garbage. They had no obvious final resting place, much like garden chemicals, old phones, light bulbs, car parts, cooking grease . . . a ton of stuff, really, now that Metzger thought about it. You weren't supposed to put any of that in the recycling bin. But you couldn't put it with the landfill-bound trash, either, although that's what many people ended up doing out of desperation or not caring or habit—or assuming (incorrectly) it would all somehow get properly sorted out by this impenetrable, mysterious entity called the waste management system.

"There's got to be a place for old batteries," Metzger assured his son. "Let's find out."

It took three phone calls to find a business near their Seattle home that would take their old batteries and ensure that they were actually recycled instead of just dumped somewhere.

Father and son decided to drive to this battery recycler so that Owen could make the delivery. On impulse, they asked a few neighbors if they had stashes of old batteries, too. Several did, so Ryan and Owen took those as well.

Owen was so delighted by this accomplishment that he and his father decided to make a regular project out of hauling one different type of problem trash every weekend to the right recycler, offering to do the same for neighbors in their Queen Anne section of Seattle. Almost any material we consign to limbo in our junk drawers, garages, cellars, and sheds holds value to someone, somewhere, Metzger figured—you just had to connect source and destination to transform someone's trash into someone else's treasure. So they started gathering bent clothes hangers one weekend, burned-out light bulbs the next, and then plastic bags, wraps, pouches, bubble wrap, and Styrofoam, none of which plays well with community recycling programs. They even started scooping up neighbors' unwanted Halloween candy after finding out about a local nonprofit, Birthday Dreams, which makes sure unhoused kids in the area get to have birthday parties.

Metzger soon realized that a lot of problem trash was accumulating out there, most of it various types of plastic that recyclers shunned, and that more people than he ever imagined were wishing someone like him would come along to take care of it. Demand kept expanding block by block as word got around about his little father-and-son project. Soon he had to create a subscriber email group to track it all, with a message going out each week on what sort of trash would be picked up next and when to leave it outside for pickup. They dubbed this "Owen's List."

Father and son were becoming full-fledged trash nerds. Owen's List was their crash course on the supply and demand of the trash and recycling industries as they became the missing link between what their neighbors thought was waste and the businesses that saw it as a valuable resource. Soon Owen's List grew a little too much. Over the holidays, they picked up close to a half ton of Styrofoam and over two hundred pounds of old Christmas lights—and he found recyclers who wanted all of it. Metzger marveled to his wife, "Who knew old Christmas lights even had a use?" (Recyclers shred them, then recover the copper and plastic for new products.)

Around this time, grateful subscribers to Owen's List who had long felt guilty about their secret trashiness started offering the duo money. A few suggested they charge for the service. "I'd gladly give up a couple lattes a month in exchange for you taking care of this," one neighbor said. "I bet a lot of people would."

Could that be true? Could their father-and-son hobby become a business that would let him leave his tech job behind and do something to help save the world? Seattle residents took pride in living in one of America's greenest cities, but would they really pay extra every month to change their trashy habits and help Owen's List patch a gaping hole in the waste and recycling system?

Metzger renamed the service Ridwell, to better explain its mission at a glance, and then set out to find out.

How Our Plastic Planet Was Born

Unlike metal, wood, clay, and glass, plastic does not occur in nature. It is a 100 percent human-made thing, a chemical concoction extracted from fossil fuels pumped from deep below the earth's surface, where they had been sequestered for hundreds of millions of years—nature's carbon capture machine. That credit card we are snacking on weekly is tangible proof that puny humans can indeed change an entire planet in a shockingly short amount of time, plasticizing the environment in a mere century. The phenomenon of plastic pollution mimics climate change—uncountable small bits of plastic that, like carbon dioxide and other heat-trapping pollutants, are local and inconsequential individually, but that, collectively swirled and spread far and wide by ocean and air currents, become a global plague that's everyone's problem. (It's worth noting that the amount of plastic we've pumped into the air, sea, and land pales in comparison to the trillions of tons of fossil fuel emissions we've released from geologic prison and spewed into the air for the last two hundred years. Just saying.)

So how did this plastic planet happen? How did we shift from an environment with zero plastic in 1924 to one in which plastic pollution became so pervasive by 2024 that we are imbibing it every day? Remarkably, this transformation took just two key events: first came the invention of a monumentally useful and durable synthetic substance; the second came decades later with the transformation of this triumph of materials science into the most insidious pollutant on the planet, tricking us all into spreading it everywhere without knowing it.

Stage one is a mostly happy fairy tale: the invention in 1907 of what was destined to become the first commercially successful, fully synthetic plastic. It took another fifteen years to perfect it, fight a patent war over it, and then bring it to market. But when it finally burst on the scene, this first unmeltable thermoplastic became one of the most revolutionary manufacturing advances in human history. And Americans knew a miracle when they saw one. Here's how *Time* magazine covered the sixty-eighth annual convention of the American Chemical Society at Cornell University in Ithaca, New York, on September 22, 1924, when the two stars of the show were celebrated chemist and inventor Dr. Leo Hendrik Baekeland and his breakout hit plastic creation, Bakelite:

> Superficially, it is a composition, born of fire and mystery, having
> the rigor and brilliance of glass, the lustre of amber from the
> Isles. . . . It will not burn. It will not melt. It is used in pipe stems,

fountain pens, billiard balls, telephone fixtures, castanets, radiator caps, etc. In liquid form, it is a varnish. Jellied, it is a glue. Those familiar with its possibilities claim that in a few years it will be embodied in every mechanical facility of modern civilization. From the time that a man brushes his teeth in the morning with a Bakelite-handled brush, until the moment when he removes his last cigarette from a Bakelite holder, extinguishes it in a Bakelite ashtray, and falls back upon a Bakelite bed, all that he touches, sees, uses, will be made of this material of a thousand purposes.

In an age defined by such transformative inventor-entrepreneurs as Edison, Tesla, and Ford, Baekeland was viewed as yet another science superhero who was transforming the world in ways so far-reaching and at such a tremendous pace that people of our era can scarcely comprehend it. US Civil War veterans were still marching in Fourth of July parades in Baekeland's time. Horses were still competing with cars, trucks, and electric trolleys to rule the streets of American cities. The electrification boom was just gearing up in urban neighborhoods, and it would take decades before rural America had significant amounts of electricity. What Baekeland offered may seem prosaic to modern sensibilities—what's more pedestrian in the twenty-first century than stuff made out of plastic? But it was seen as a magical material then, boundless in its potential, freeing civilization from the tyranny of nature and the limits of old-world materials.

Time's coverage of chemists gathered in Ithaca to honor Baekeland and ponder plastic world domination might now seem over the top, but it was in fact eerily prescient. Just substitute the word "plastic" for "Bakelite" to account for the thousands of different plastic formulations that followed Baekeland's creation, and every prediction in the article not only came true but turned out to be understated. Bakelite's most important use was in a host of industrial and infrastructure systems, including key components for the rapidly expanding electrical grid. But it's best remembered for its use in fashioning beautiful products, boldly colored with art deco designs: Bakelite clocks, jewelry, sculptures, fancy buttons, telephones, lamps, and iconic living room radio sets (first-gen entertainment centers that are prized by contemporary collectors). Streamlined house trailers and campers were built out of this gleaming material, looking like early prototypes for spaceships. Earlier plastics had replaced the extinction-driving harvest of ivory and tortoiseshell for billiards, pianos, combs, and eyeglasses. Plastic celluloid, the very first plastic, made movies possible, and another form of early plastic insulated thousands of miles of electrical cables for the transoceanic telegraph and telephone lines—the first World Wide Web. But Bakelite made plastic ubiquitous—and put the plastic industry on a trajectory to become the titan of wealth and pollution it is today, worth $1.2 trillion.

Baekeland got the full hero treatment from *Time*, including the requisite cover, featuring a dour headshot of the balding, Belgian-born chemist. The magazine portrayed him as a shy man of science more than a business tycoon, "a man in middle years, erect, rugged, taciturn," and, rather oddly, "with the sensitive mouth of a field marshal and the cold eyes of a philanthropist." He was, however, a more complicated character than the magazine suggested, an inventor whose earliest breakthrough was creating something very different: Velox, the first commercially successful photographic paper. He and his two business partners sold their business and exclusive rights to the revolutionary material in 1899 for what was then the fabulous sum of $750,000 to Eastman Kodak, then in the process of building its own pre-digital photography empire. Baekeland was in his mid-thirties at the time, and his share of the sale was worth about $8 million in 2023 dollars.

There was a condition to the deal, though: a noncompete clause, which meant no more photography-related inventions for Baekeland. So he decided to try his hand in the then-nascent field of polymer research. When asked years later why he chose a branch of chemistry he knew little about, Baekeland replied: to make money. And so he did, and a whole lot more. That noncompete clause changed the world.

He set out to invent a synthetic shellac, a protective finish for wood made from the waxy secretions of the female lac bug. Instead, out of his beaker came Bakelite. For that, he would be dubbed the father of the plastics industry. He is said to have popularized the term "plastic," too, from the ancient Greek *plastikos*—capable of being molded or shaped.

Baekeland sold his company in 1939 to the chemical giant Union Carbide, then best known for its Prestone antifreeze. That company in turn was eventually gobbled up by Dow Chemical, a major chemical and plastics maker. Baekeland retired to his winter estate in Coconut Grove, Florida, where he obsessively tended his huge tropical garden. Increasingly eccentric and reclusive in later life, he would eat food only if it had been stored in cans—never plastic—until his death of a stroke in 1944 at age eighty. Maybe the man who once said his surname translated into English as "land of beacons" sensed something about the dangers of eating plastic even back then.

If so, it didn't slow down the plastic juggernaut he set in motion. Modern life has been built on the back of the shapeshifting materials Baekeland and his successors developed. Without plastic, few of our most transformative technologies and products, for good or ill, could have gotten off the ground as we know them. First, it provided the combat boot soles, tank treads, radio sets, and thousands of other military tools, technologies, and weapons of World War II. Then came the invention of the slippery plastic coating later known as Teflon (along with its by-product, PFAS forever chemicals) in connection with the Manhattan Project's quest for the first atomic bomb. After the war, the stockpile of chemicals and manufacturing capacity built for the military were channeled into a plethora of civilian products: circuit boards and computer chips, high-speed data transmission, the internet, smartphones, spaceships and satellites, joint replacements, pacemakers, lifesaving arterial stents, and a host of other medical devices. Astronaut pressure suits, hazmat suits, wet suits, fiberglass boat hulls, surfboards, surf fins, surf wear—all require plastic, lots of plastic. Modern cars are half plastic by volume. There'd be no easy way to make renewable energy with solar panels or wind turbines without plastics. The list of plastic-dependent breakthroughs and everyday items is nearly endless. Just count each time you touch something that's made of plastic during a typical day and you'll see.

One quality that especially excited the engineers and product designers of Baekeland's time was plastic's near invulnerability. Certainly it can weather, wear down, or break through the application of force like any other substance. But plastic does not rust, corrode, or tarnish like metals. It does not shatter like brittle glass or ceramic. Wood termites don't bore into it. It resists rot from bacteria or mold, as virtually no microorganisms can digest, diminish, or be nourished by it. The tireless forces of decomposition and decay that prey on natural materials hold little sway over the synthetic power of plastic. No one had seen anything like it before. Chemistry had magicked something entirely *unnatural* into the world.

A Useful Life Measured in Minutes, a Waste Life Measured in Centuries

What the designers, manufacturers, and marketers didn't think about in 1924, and still are refusing to deal with fully today, is what happens to such invulnerable stuff when we're done with it—when it finally wears out or breaks or otherwise gets thrown away. This was a problem from the very start with durable plastic products, which were designed with their *use* in mind, not their *demise*. But that became a catastrophe and a curse when greed and folly led to that second part of the story of plastic, the innovation that turned the fairy tale into a nightmare: the creation of *disposable* plastics.

Suddenly, instead of durable plastic products that lasted for years or decades, we had much smaller bits of instant trash—bottles, packaging, containers, wraps, and other disposables—that nature can't process or absorb, and that humans can rarely manage to recycle (despite industry claims to the contrary). A patently insane proposition—flooding stores, homes, offices, and schools with products that have a useful life measured in minutes and a waste life measured in centuries—was marketed as a convenience, as a cost savings, as a feature, not a bug, of modern life. Why wash that cup when you can just throw it away? Why inconvenience your customers by having them lug glass bottles back to the store (which had been common practice in the United States since the 1700s) when you can emboss them with "no deposit, no return," and tell people to just toss them and—this was a crucial part of the pitch—never worry about them again? A deluge of print and television ads in the 1950s, '60s, and '70s promoting this waste-more, want-not economy literally trained consumers to embrace disposability. Typical was one magazine ad featuring dual images: a frumpy, frowning person struggling with a bursting bag full of deposit bottles, contrasted with a smiling fashion plate tossing her containers, each with a plastic straw sticking out of it, into the trash can. "Tote those empties?" the ad asks. "Or toss them away?" Another ad asks: "In this disposable age, is there a reason for the non-disposable bottle?"

This disposable age. That phrase says it all: the branding of waste not as a problem but as a way of life—an idea that went from strange and possibly wicked, to a wasteful but acceptable trade-off, then finally to desirable and inevitable, a habit we can scarcely imagine living without. And starting in the 1970s, with the invention of the first plastic soda bottle that didn't burst from the pressure of carbonation, disposable plastic demand and profits exploded.

Why would the chemical and plastics industry do such a thing? Simple: it doubled the market for plastics practically overnight, not by making beautiful, long-lasting products that require teams of engineers and designers, but by making instant trash. Plastic packaging and containers now account for half of all plastics in the world.

The effect of this was dramatic. By the 1990s, plastic waste had tripled. And that was just the start of the acceleration: the first decade of the 2000s produced more plastic waste than the previous forty years combined.

By 2010, a little more than a century after Bakelite's invention, and about sixty years after the dawn of this disposable age, the world was churning out 300 million tons of plastic waste a year, which is why it's now *everywhere.* Almost every piece of plastic ever made still exists in some form. With only 9 percent of plastic trash recycled and, at best, about two-thirds safely ensconced in landfills, that still leaves a lot of plastic waste to find its way into oceans, rivers, soil, animals, and us.

And the tide keeps rising. Estimates for 2023 put worldwide annual plastic waste at more than 400 million tons a year, with the United States contributing the largest share of any country. Every day, it is literally snowing microplastics from the point of view of fish in the sea.

So how do we even visualize such a staggering amount of plastic waste? Well, 400 million tons is roughly the weight of three thousand modern cruise ships—not counting the weight of the four thousand or more passengers each can carry. That's ten times the number of all the cruise ships in the world. Or if you prefer a dry-land comparison, it's equal to the weight of 1,100 Empire State Buildings, which is roughly six Manhattan skylines' worth of skyscrapers (defined as buildings over three hundred feet tall).

But even those comparisons fail to convey the true magnitude of 400 million tons of plastic waste, because ships and buildings are very heavy, dense objects of metal and concrete, while plastic is very light, so it takes much, much more of it to get to the same weight—which means it takes up a lot more space than 1,100 high-rises. So let's visualize the *volume* of 400 million tons of plastic, too. Imagine for a moment that, instead of all the various shapes and sizes of plastic products, this waste consists entirely of the familiar 16-ounce disposable water, soda, and sports drink bottles, which the world consumes at a rate of about a million a minute. The volume of 400 million tons of them would fill 60 percent of the theoretical capacity of America's largest reservoir, Lake Mead, the 247-square-mile, 1,229-foot-deep body of water behind Hoover Dam. That amount of plastic waste represents more than double the amount of water actually in Lake Mead as of 2023. Far more than three thousand cruise ships could float on that sea of beverage bottles—it would actually accommodate more than fifty thousand of them. That's how much plastic the world throws away in just one year. With America leading the pack.

At current rates of global plastic-production growth, this amount of waste will more than triple by 2050—nearly two Lake Meads a year, filled to the brim. Soon the accumulated mass of the portion of plastic waste that finds its way into the oceans will outweigh all the sea life in the world if we don't change course. And there it will stay, for out in the wild, disposable plastic remains plastic, as sun, wind, storms, wave action, and other physical forces only break it up into ever smaller pieces. It doesn't become fertilizer or soil or food for living things, as natural materials do. Add to that the prodigious stream of microplastics from tire wear and washing machines and, in a recently discovered nasty surprise, plastic pollution generated by the recycling process itself, and it becomes obvious how plastic has infested our soil, air, water, food, and bodies.

How Do We Rid Well?

To walk through the Ridwell warehouse in Seattle's south of downtown district is to take a grand tour of the father of the plastic industry's unintended legacy: a disposable, single-use economy made of zombie trash that will not die.

The big room with the high ceiling and crammed aisles jars the senses with its piles, boxes, pallets, and bags of waste. It looks as if a landfill has been excavated, then its contents sorted, bundled, and neatly organized. That's not far from reality, except this material has been rescued *before* its more typical destiny as landfill fodder, litter, or waterway pollution. And there is a lot of it: this "stock" changes day-to-day, the tide that never stops, with most of the warehouse contents turning over every two or three days.

Ryan Metzger, his hair close-cropped, wearing a Ridwell-branded, construction-worker-orange T-shirt, navigates the crowded, noisy warehouse floor like an experienced rock climber picking out a path to the summit. One minute he's scanning the housewares aisle for fresh arrivals, next he's surveying the packing of a battery barrel with flame retardant and sand as it's readied for transport, and then he's looking over a shipment of plastic film and bags being loaded by forklift into a trailer backed to the dock. Soon it will be headed out to a company that sees this otherwise unrecyclable trash as raw material for new products.

One area of the concrete floor is dominated by a mountain of giant translucent orange bags filled with Styrofoam blocks, chips, cups, clamshells, peanuts, and plates. Nearby are compressed bales of plastic film, wraps, bags, and tens of thousands of those distinctive blue and white Amazon mailing pouches. Other piles wait their turn in the baler. Another area contains those impossible-to-recycle composite pouches made of multiple layers of materials—for chips, dog food, frozen food, granola bars, candy bars, tuna, and even (sigh) organic foods—that you see all the time. These are often distinguished by their silvery lining inside an outwardly normal-looking plastic food bag. Most people believe these are recyclable. They are not. Near these composite plastics are boxes filled with bread tags—the ubiquitous small, flat, plastic slotted squares that seal the plastic bags for bread, too small for almost all community recycling centers, as they would slip through or clog the machinery.

Ridwell has come a long way since the father-and-son Owen's List days. The question Metzger posed back in 2018 about its viability as a business—would enough people be willing to pay extra every month to have their zombie trash hauled off and turned into sustainable products?—has been answered emphatically: yes. With an annual growth rate of 50 percent and more than one hundred thousand subscribers as of 2023, Metzger is still occasionally stunned by Ridwell's success. Sure, he says, Seattle and the Puget Sound area, with its strong civic and popular commitment to sustainability, seemed like an ideal test case for Ridwell. But even on such fertile ground, persuading a critical mass of people to pay $14 to $18 a month on top of their city utility and waste bills was a big ask. Yet in some of its most successful zip codes, Ridwell was soon getting 15 to 20 percent of households to sign up, a level of uptake that made it clear Ridwell could do well.

The company now has more than two hundred employees and a fleet of colorful vans (the latest is an EV) that make pickups every two weeks. The trash they gather has been sorted by customers into Ridwell cloth bags labeled for each of the six most common categories of waste that community recyclers can't handle: "plastic film," "multilayer plastic,"

"plastic clamshell containers," "batteries," "light bulbs," and "threads" (for clothes, shoes, and other textiles). For each pickup, a different "featured item"—an unrecyclable type of trash that doesn't accumulate quickly enough for twice-monthly pickups—is also specified by email. It might be corks one time, small electronics the next, then nonperishable food, holiday lights, cords and chargers, pill bottles, bread tags, plastic bottle caps, or unused diapers (left over when a baby graduates to larger sizes or big-kid underwear). Each customer also gets a white metal porch bin emblazoned with Ridwell's distinctive orange script to fill for pickup days—as well as to serve as highly visible neighborhood advertising.

After a year of expansion within the Seattle city limits, Metzger and the other three founding partners—all original Owen's Listers—started seeking subscribers in surrounding towns. Soon Ridwell's service area stretched 90 miles north to Bellingham, Washington, through Everett and Snohomish County north of Seattle, and 60 miles south to the state capital, Olympia. Ridwell's next logical outpost was just across the state line, Portland, Oregon—also perennially on the list of greenest American cities—followed by metropolitan areas in seven states, including Minneapolis–Saint Paul, Denver, Austin, Atlanta, Santa Monica, and the San Francisco Bay Area. Los Angeles is next in the company's expansion plan.

Six years in, Metzger chalks up Ridwell's success as partly due to the convenience—the company makes it very easy for people to get problem waste taken care of—and partly due to the fact that recycling is something his subscribers tell him they care about deeply. Ridwell provides a straightforward and relatively low-cost, feel-good way for customers to embrace the idea that individual and community actions can have an impact on repairing environmental and social ills.

One more factor helped Ridwell's successful launch: it benefited from phenomenally fortuitous timing. The company's start-up moment coincided with lawsuits against beverage bottlers, legislative hearings on plastic waste, and a changing waste-export market that exposed systemic failures and frauds throughout the US recycling business. And right on the heels of those damaging revelations, the entire recycling supply chain ground to a halt in the wake of COVID-19, with many communities just throwing up their hands and diverting everything to landfills.

Recycling had already become a black hole to many Americans—exactly what could be recycled, and what was done with our stuff after we left it at the curb, was mysterious rather than clear-cut. Then Ridwell arrived to fill that information and trust gap by making transparency a core part of the business. The promise was explicit: the Ridwell website would tell you exactly where your stuff is going, what's happening with it, what products the material will be turned into, and how it benefits the community and environment. No one had ever seen anything like this, not even in green Seattle, where you chucked stuff into the bin and never heard from or about it again. Suddenly, at least in this little corner of the waste universe, it was possible to see what recycling could and should look like.

Ridwell's leading collection item is plastic film, a category that includes wraps, bags, bubble wrap, cereal box liners, shipping air pillows, and grocery bags—all of which get sent to a Virginia-based company, Trex. The company got its start when inventor Roger Wittenberg figured out how to build a nearly indestructible outdoor bench out of shredded plastic bags and sawdust in 1988. Later, as Wittenberg perfected the process and expanded Trex's range of products, the company was bought by Mobil Oil, then later spun off as a stand-alone business when Mobil and Exxon merged. Trex is now a billion-dollar company that turns plastic film into low-maintenance decking, furniture, and other outdoor structures normally made out of wood. Though its "boards" look like wood, they are made out of 95 percent recycled plastic film. Trex guarantees that they won't wear out for at least twenty-five years, during which time the stuff requires no painting or varnishing, and it won't rot, warp, fade, or get eaten by termites. The company's pitch is that the greater purchase price for its recycled products compared to wood is worth the investment because the durability and low maintenance make the overall cost of ownership over time much less. The goal is to save trees and energy while keeping a pernicious form of unrecyclable plastic out of the environment.

Trex will basically take everything Ridwell (and anyone else) collects and has so far diverted 400 million pounds of plastic film.

Another popular item for Ridwell pickup are those ubiquitous pouches that combine different plastic polymers and sometimes aluminum into an unrecyclable amalgam. These laminates are almost always landfilled, though many people toss them into recycling bins by mistake. Separating the different materials bonded together to make this packaging is technically challenging, and the end result would be of little value, and certainly far less than the cost of separation. You couldn't design something less recyclable if you tried. In recycling industry lingo, such merging of materials is considered "contamination," which renders the material useless. That's the reason "paper" coffee cups, which have a polyethylene coating on the inside to keep the coffee from bleeding through the paper, are rarely recycled: because, you guessed it, the paper is *contaminated* with plastic. On purpose. And we Americans use about 54 billion disposable "paper" cups a year (that's all types, including coffee, soda, smoothies, ball game beers, etc.), which is an astonishing 160 disposable cups for every man, woman, and child, all of it unrecycled because of the contamination problem.

But Ridwell found multiple solutions for those pesky contaminated composite pouches. Some go to a company based in Santa Ana, California, named Arqlite, which turns them into "smart" gravel that can be used for gardens, hydroponics, roofing, drainage, and other purposes, in place of natural gravel and manufactured crushed stone. Such plastic products often raise concerns that they can emit chemicals into the environment, but the company claims extensive testing has shown the product is "inert and stable" with no leaching.

The other company that takes unrecyclable plastics of this sort, ByFusion, in Los Angeles, uses steam and pressure to mold plastic into construction blocks that fit together like giant Legos and have a wild, multicolored surface—reminiscent of a trashy Jackson Pollock knockoff. These rectangular cubes are designed to take the place of the venerable concrete block—with 83 percent lower carbon emissions. Concrete manufacturing, the world's number one product, generates massive amounts of heat-trapping pollutants, about 8 percent of all human-made carbon emissions. So ByFusion is attacking another big problem in addition to plastic pollution. It sells a machine that spits out these blocks right at construction sites. As long as there is plastic waste nearby for raw material—which is basically anywhere in the world—there is hardly any transportation footprint or emissions for this concrete substitute.

Ridwell's solution for another problem plastic, Styrofoam, is sending it to one of several companies around the country that use machines called densifiers, which compress the toxic, crumbly foam plastic into a dense, hard, durable plastic. It's used for picture frames, home-decor molding, and an array of other products, including material that can be used as an additive to lower the carbon footprint of cement.

Unwanted clothing—85 percent of which ends up in the landfill in the United States—is transferred by Ridwell to a variety of reuse and recycling destinations, from nonprofits such as Goodwill and other thrift outlets, to charities that supply clothing to children in need, to cloth recyclers that turn old duds into reusable cloth wipes. Another company, ReFleece, takes worn-out outdoor and performance wear from Ridwell and converts the material into smaller items—pouches, purses, reusable bags, and laptop sleeves. Though we don't usually think of what we wear as a source of plastic waste, the overwhelming majority of the clothing we trash is made partly or entirely of plastic. Even "all cotton" clothes have plastic waistbands, plastic buttons, or other plastic components.

In all, Ridwell and its customers have kept 15 million pounds of their discards from being landfilled so far. The company gives away, sells and, with particularly difficult materials, pays recipients to take the stuff, but only after vetting each one to verify its sustainability bona fides. This is an exceedingly rare practice in the normally dump-first, ask-questions-later business of waste management, where transparency is the exception, not the rule. Ridwell's tracking shows that 97 percent of the discarded material it collects is reused or recycled.

Metzger describes Ridwell's subscribers as a community rather than customers, part of a movement "to build a future without waste . . . taking small steps that add up to big change." The company posts stories on its website about this community, focusing on individuals' waste-reduction achievements as well as their ideas for items that Ridwell hadn't considered before. One Seattle resident suggested the collection of used eyeglasses, which Ridwell now picks up periodically and supplies to a nonprofit that repairs and cleans them, then matches them to unhoused and low-income families who can't afford new glasses and have been forced to struggle with poor vision. Other users in Portland connected Ridwell to a nonprofit called the Crayon Initiative, which supplies free crayons to kids in hospitals. Old, broken crayons and stubs are melted down and formed into new ones, then given to young patients to brighten their hospital stays.

As a rescue operation for the zombie trash of our disposable age, Ridwell and its community of waste-cutters are both inspiration and paradox. With its origin story of a father-son recycling project that spread neighborhood by neighborhood, it would be hard to imagine a better demonstration of the powerful impact that individual and community choices can have. By organizing as a social impact company, Ridwell has been able to attract customers and investors who value enterprises where the return on investment isn't just financial but also the fulfillment of the core mission. And in this case, that mission is not only to reduce waste, plastic pollution, and carbon emissions but to set an example of how smart waste management can also support the reuse economy and effective, transparent recycling.

For the Rest of Us

That's the positive part—the Ridwell community really is making a difference on the zero-waste front. But (and there's a big *but*) Ridwell and the few other companies doing similar work are also a constant reminder that the recycling system everywhere else is total garbage. Ridwell operates in only a handful of cities, after all, and at best it captures only 15 to 20 percent of the neighborhoods it serves. So even in those places, and no matter how much Ridwell might spread to other communities, 80 percent of this perfectly useful, reusable, and reclaimable material is still going to the landfill or becoming pollution. Outside of the cities Ridwell and like-minded companies serve—the most successful being a similar enterprise in Philadelphia called Rabbit Recycling—100 percent of this treasure is trashed.

Even though there are proven technologies for making useful products out of plastic film waste, most of the country isn't bothering. As a social impact company, Ridwell doesn't worry about making a high profit from collecting wraps and bags to send to companies like Trex. But the recycling industry overall is very much profit-driven, and manufacturing new polyethylene, the stuff plastic film is made from, is far cheaper than recycling and repurposing it, given the perverse economics of the disposable economy, in which the waste creators bear no financial consequences for its cleanup costs. In the end, less than 2 percent of the 9 billion pounds of bags and wraps that Americans throw out each year gets recycled.

This is true, to one degree or another, for all plastic containers and packaging. This is why determining which products cause the most problems and where is so crucial, though figuring that out and getting rid of those materials is like trying to solve a crime with no witnesses. This has stymied efforts to stem the flow of the worst forms of plastic waste and fix the recycling system—the data is paltry at the global level, and nearly non-existent when it comes to figuring out how to counter the wasteful disposable economy at the local level. Ryan Metzger likes to say he can't wait for someone to crack the code on that problem, even though ending the scourge of plastic waste would put a big dent in Ridwell's business. "Nothing would make me happier," he says.

Across the country, an engineer based in Georgia, who has watched Ridwell's move into Atlanta with interest, is hoping to make Metzger's dream come true by helping the world figure out how to slay plastic pollution at last, one community at a time.

2

Trash Genius

For someone who suffers dizzying sensory overload in grocery stores, gets headaches from the laundry soap aisle's cloying perfume, and long ago ceded food shopping to her husband, Jenna Jambeck sure spends a lot of time in supermarkets.

"That's where the data is," she says with a sigh. "If you want to understand plastic waste, you have to go to the source."

Such is life for the world's only official Trash Genius. A University of Georgia environmental engineering professor and a recipient of a MacArthur "genius grant" for her groundbreaking work on plastic pollution, Jambeck may hate shopping. But she *loves* data.

We are a few steps inside the Tifton, Georgia, Walmart Supercenter, where Jambeck pauses to get her bearings before searching the aisles for data on the plastic packaging entering this rural community of seventeen thousand. She already spent a dreary autumn morning surveying the litter along a hundred-meter stretch of roadway through a nearby neighborhood. The detailed cataloging and photographing of the area felt like a crime scene sweep, though here the evidence consisted of empty soda bottles, crumpled chip bags, and ketchuppy fast-food remains instead of fingerprints and murder weapons. These litter surveys show what kinds of plastic packaging waste are flowing *out* of the community and becoming pollution. Jambeck calls the whole process CAP, for Circularity Assessment Protocol. Even the packaging for her take-out meal the night before became data.

"Do you see the rice anywhere?" she asks, peering at the overhead signs that stretch into the distance of the cavernous building, a supermarket embedded in a superstore. She grimaces at the noisy sprawl of the place, the sale items stacked in towers around the entrance, Bonnie Raitt's "Right Down the Line" blaring from the sound system. There are approximately three thousand other places she'd rather be this morning, Jambeck says, first among them the local landfill, a word she pronounces with the happy inflection kids use for "Disneyland."

But this is fieldwork, not shopping, the only imperative that can keep her in a supermarket for hours rather than minutes. "That's why my husband does the shopping. Because when I go, I'm in such a rush to leave, I just grab the three things I like—popcorn, bubbly water, and maybe some chocolate—and run out of there. Then I get home and my family looks in the bag and says, 'Okay, but what's for dinner?'"

We start roaming until we spot a likely location for the rice several aisles away and tromp off to see what Rice-A-Roni and Ben's Original and their many competitors can tell us about plastic pollution. She will make a detailed analysis of the packaging for all the rice brands on the shelves: what it's made of, whether it's recyclable, if it's labeled with clear instructions to consumers on what to do with it when the contents are used up. You'd think manufacturers would make basic packaging information about their products publicly accessible to their customers. But no, the disposable economy is an industrial black hole—one Jambeck is illuminating by building a shareable packaging database from her many grocery store and litter surveys. The information gathered here will be used both for a local plastic pollution plan in Tifton and in her global packaging project supporting researchers and plastic waste strategies worldwide.

She's been helping communities counter plastic pollution for years, from villages in Vietnam and India to towns and cities like Tifton and Miami. Her global data on plastic pollution galvanized the world in 2015, when she and her fellow researchers quantified for the first time how much plastic is dumped in the oceans yearly: nearly 9 million tons, far more than anyone guessed. That data point and the research that followed made her an international science superstar, with a platform and audiences that range from middle school beach-cleanup volunteers to world leaders. Not that the soft-spoken Jambeck would ever describe herself that way. Her mental mirror still reflects the kid from tiny Pine City, Minnesota, where there was no curbside trash service and her favorite chore was driving the family garbage to the town dump with her mom. This was no fancy landfill, just an open pit on the edge of a wooded area. They'd borrow a neighbor's truck, fill it with their trash, back it up to the edge of the pit, and throw it in—a dry lake of garbage that endlessly fascinated young Jenna with its ceaseless variety and mystery. She especially liked it when they paused to watch the bears that emerged from the woods to forage for food among the waste.

We have reached the rice aisle. Rice is one of the packaged staples she always checks, along with sugar, cooking oil, eggs, milk, and laundry detergent, wrapping up with the array of single-serve snacks in the checkout area—little bags of chips, candy, pretzels, and such, which tend to be more packaging than product. The rice occupies a twenty-foot-long section of the aisle, shelf upon shelf holding a multitude of multicolored packages in every form and material imaginable. Jambeck begins to take notes and photos of this rainbow of container variety.

"I should warn you," Jambeck says. "People who do this with me sometimes have, um . . . strong reactions. Sometimes they say, 'I can't believe I never noticed that before,' or 'I wish I could unsee that, but I can't,' or 'Thanks a lot! Now I can't go to a supermarket without crying.'"

She pauses in her photography, then adds, "Well, maybe not *actually* crying, like bursting into tears. I don't want to exaggerate. I think it's more like they *want* to cry every time they go shopping."

And then she shows me why.

Some packages are marked as recyclable. Others qualify that claim with a warning that they must be taken to a proper recycler, without further explanation. Some packages have a big slash mark through the recycling symbol, while quite a few other labels are silent on the question of recyclability. All contain the same basic product, the only differentiator being the packaging. There are pouches, boxes, boxes with plastic windows, single-serve cups, plastic jars, composites of cardboard, foil, and plastic, clear plastic bags, opaque plastic bags, and sacks that are burlap on the outside but plastic on the inside. Little of this is standardized; there is great variation in shape, design, and materials, all intended to stand out on the shelf as unique—which is the one packaging strategy guaranteed to make recycling as difficult as possible. Companies call this variety and consumer choice, although dozens of these seemingly separate brands are actually owned by the same three or four parent companies that dominate the rice industry. There are many varieties of rice, but in most American supermarkets, the main choice seems less about the product—which for most shoppers comes down to choosing white or brown, regular or organic—and more about the packaging. And the packaging often seems completely arbitrary.

Australian comic Jimmy Rees does a bit called "The Guy Who Decides Packaging" that crystallizes this reality. The Guy is sucking down martinis at an alarming rate and deciding on a whim what product goes into what container and giggling drunkenly after each choice. Milk? If it's small, put it in a box. If it's large, put it in a plastic bottle with a handle. Orange juice? Put it in a different plastic bottle with no handle. Potatoes? Put them in a plastic bag. Oranges? Put them in a plastic net. Tuna? Put it in a can with a pull tab. Beetroot? Put it in a can with no pull tab. Plastic ziplock bags? Put them in a box. Paper bags? Put them in a plastic bag. He goes on in this vein, the absurdity mounting. It's funny because it's true, though seeing the reality so starkly in the rice aisle is not amusing at all.

"So here's a question for you about these choices," Jambeck says. "Let's say you are trying to find the healthiest rice you can buy here. And you want it in the most recyclable packaging or, at least, a package made with the least amount of plastic. Which would you choose?"

We stare at the eye-watering crazy quilt of rice packaging and ponder this question, which seems almost impossible to answer. The healthiest and organic products are all in plastic. The most expensive per ounce and the least healthy, processed instant, and salt-laden flavored rices are the only ones in readily recyclable cardboard. The recycling guidance on many of the labels is impenetrable or not there at all. The packaging is distinctive, even eye-catching, but not particularly informative of the product inside. Rice, after all, is a simple commodity that could easily be sold as a bulk product out of big bins and that customers could bring home waste-free in their own reusable containers. It would be efficient to do it that way. It would be cheap. It was the way of things for most of human history. It would eliminate much of the scourge of disposable packaging. And there are stores devoted to this method of buying staples—grains, coffee, and other bulk products—but zero-waste markets are few and hard to find, particularly in the United States. Walmart, which sells 40 percent of the groceries in the country, is definitely not one of them.

We move throughout the store and repeat the same exercise with similar results for sugar, oil, and detergent. Even milk and eggs have surprising packaging variety, although the traditional glass bottle and cardboard egg cartons remain the only remotely sustainable choices. Recyclable cardboard milk cartons are still around, but many brands have replaced the ingenious "gable top" built-in pinch-and-pull spout that's been around since 1915 with those little round, unrecyclable plastic pull tabs. Aisle after aisle, it's the same story—packaging choices that frustrate recycling and add nothing for consumers. The soda aisle is not part of Jambeck's survey, but we walk down it anyway. Soft drinks command more shelf space than anything else in this grocery store—a product with no nutritional value and quite a lot of negative health effects. Cans and bottles and boxes of sugar chemical water are stacked there in a profusion of different products and brands, though they, once again, are primarily manufactured by the same three or four companies, with not a reusable container in sight. And unless this aisle is in one of the ten US states with a "bottle bill" that provides an incentive for recycling in the form of a small deposit on beverage containers, very few will be recycled.

If you shop like I do, you are not really seeing the absurdity of this day-to-day. We just focus on our shopping lists, on finding those items we need, and getting the heck out of there as fast as we can. We either toss stuff in the basket out of habit, because that's the kind we always buy, or we find the lowest-priced version and grab that one, regardless of brand or packaging. Is that the healthiest or most sustainable? Who knows!

"Yes!" Jambeck says. "What's being sold here isn't consumer choice— it's consumer confusion. I basically have a PhD in trash, and even I can't figure out what's the healthiest and most recyclable product."

And then Jambeck's point, the thing that people who accompany her to the market can't unsee and that makes them want to weep, suddenly becomes clear. This illusion of variety is really just a proliferation of future plastic pollution. What can't be unseen is that this vast array of the same basic product in different, rarely recycled packaging exists not because it's beneficial to consumers, but because it's so much more profitable for the companies to make it this way—so long as they don't have to bear the cost of dealing with the plastic pollution crisis they create. The entire food industry is organized around this business model. What they are really selling in every aisle in the supermarket is the waste. The food is just along for the ride.

Before I can stop myself, I echo one of the common responses Jambeck gets from those who accompany her on these surveys: "I never really noticed this before. I never thought about it this way." She just nods, welcoming me to the realization that she had years ago: that the challenge of slowing this, changing this, fixing this is monumental. I say, "This must drive you crazy."

She laughs. "Why do you think my husband has to do the shopping?"

The Dirty Truth about Recycling

For years, many waste experts, sustainability champions, and leaders of some of our greenest cities laid plans that bet the future on recycling— that, basically, we could recycle our way out of trashiness if we just pushed hard enough. Divert everything from landfills was the mantra; recycle and compost everything was the goal. We were on a race to zero waste, and that sounded sensible, even awesome. But it wasn't.

What most of us didn't know then (and many still don't realize now) was that for nearly three decades, our recycling bins contained a dirty secret: half the plastic and paper we put into them never made it to our local recycling centers. Instead, it was stuffed onto giant container ships and sold to China.

Fourteen thousand or more nautical miles later, all those bales of mixed plastics and paper, much of it the sort of hybrid packaging on display in the average supermarket aisle, were as likely to be burned or dumped, or become ocean plastic pollution, as recycled. And the half that *did* get recycled underwent processing with such lax environmental contols, it would be illegal in the United States. Yet everything was counted as "recycled" in our record books.

Our recycling system, in short, was beyond total garbage. It was also a lie.

This mass export of recyclables began around 1992, when corporate America was in the midst of an orgy of outsourcing our manufacturing prowess overseas. In a quest for lower wages, lax labor regulations, and even laxer environmental oversight, the US private sector offshored irreplaceable jobs, knowledge, and industrial capacity so that China, among others, could make everything from our computers to our underwear to the critical components of cars, trucks, electronics, infrastructure, and even our national defense. Desperate for raw materials, Chinese companies were only too happy to buy our garbage, too, using it to make packaging for the products they were selling back to us.

They paid top dollar for even the most contaminated, least valuable plastic and paper waste. In the States, this stuff would be rejected out of hand by recyclers and manufacturers. It would have to be carefully sorted first, the unrecyclable plastics rooted out, and the recyclables separated into different types of plastic and cleaned—an expense of time and labor. But China took it all, as is, dirty and unsorted. The private waste management companies that dominate garbage collection in the United States enthusiastically embraced this as a less-work-for-more-money strategy. Meanwhile, the weekly recycling we hauled to the curb, thinking we were doing something good for the world, half the time ended up as mountains of stinking litter and toxic waste in China's impoverished, unsafe, and unhealthy "recycling villages," the arrival point for our garbage exports.

Thousands of these polluted shantytowns on the edges of big cities contained mom-and-pop plastic recycling businesses, reeking of caustic chemicals and open-pit garbage burning. An estimated 60 percent of American and European paper and plastic waste ended up in such places. Workers sorted, extruded, cleaned, and liquefied this plastic trash, dumping the useless materials and residue into pits and polluted canals with banks encrusted with plastic debris.

Jenna Jambeck has visited these villages and laments their dire conditions and lack of pollution containment. "They were given no infrastructure to work in or to manage any discards from the recycling," she says. "Materials were dumped and, inevitably, found their way into waterways and the ocean."

Meanwhile, America's once robust capability to sort, clean, and recycle its own waste decayed. Why invest in expensive technology and labor to keep up with the constantly changing world of packaging and plastics when the mess could be bundled off to China in exchange for easy money and the appearance of being green? In just a few years, US recycling became completely dependent on China, as trash became the largest US export by volume.

Then, in 2018, the sham ended. After years of warning the United States and other countries that it was going to end imports of dirty foreign garbage no one else would touch, China finally pulled the plug. And American recycling collapsed.

Waste companies, cities, and towns accustomed to premium prices for poor-quality recyclables were forced to raise collection fees, reduce the types of plastic they'd accept to one or two, or simply end recycling programs and send it all to landfills. What had been a revenue source suddenly became a massive expense that staggered towns and cities across America: Stamford, Connecticut, went from earning $95,000 from its recyclables in 2017 to losing $700,000 in 2018. Seventeen miles away, Fairfield went from a profit of $64,000 to a cost of more than $250,000. Prince George's County, Maryland, went from earning $750,000 to losing $2.7 million. And Bakersfield, California, swung from earning $25 a ton for its combined unsorted recyclables (glass, plastic, paper, metal) to *paying* $70 a ton just to get rid of it. When every American alive is averaging a ton and a half of trash yearly, such deficits add up fast.

Although Chinese leaders spoke of their decision to stop serving as the world's trash compactor as part of a larger domestic crackdown on pollution and poor waste management, the move was widely depicted in the United States as an act of near aggression. It probably didn't help that the Chinese name for the crackdown translated into English as "National Sword." Industry folks accused China of bringing about the death of recycling, but the hard truth was that National Sword did not cause a crisis— it exposed one that had been there all along.

Jambeck, among others, saw China's move as forcing a long-overdue day of reckoning—one that might even be beneficial in the long run, a catalyst either to make packaging more recyclable or to shift America toward reusables.

"China . . . did us a favor," she said at the time. "They ripped off the bandage that had kept us from seeing just how bad things were."

But a little more than a year later, the pandemic hit, and the still-reeling recycling system took another body blow as global supply chains were disrupted. Recycling rates plunged further. More curbside collection programs were cut back or cut entirely. By 2023, recycling had made a partial comeback, but was still limping along, with recycling rates down and landfilling up—even as packaging waste and disposable plastics hit record highs in the United States and worldwide.

None of this means we should give up on recycling—environmentalists and packaging industry advocates can agree on that much. Parts of the system work well, while others scrape by but need improvement. We're good at recycling paper and really, really good at recycling cardboard boxes—more than nine out of ten are recycled in the United States. Metal is in the we-could-do-better category. Aluminum is, like glass, infinitely recyclable, and though recycling both materials saves energy and reduces pollution, aluminum is a superstar. Using recycled aluminum saves massive amounts of energy and water while avoiding strip-mining and the release of toxic pollutants and heat-trapping emissions associated with virgin aluminum. There's absolutely no comparison: recycling aluminum uses only 5 percent of the energy needed to refine new aluminum. And yet we still manage to recycle only about half of all aluminum beverage cans, and far less of the total aluminum discarded in foils, packaging, old appliances, and other scrap—only about 17 percent of the 4 million tons of aluminum discarded yearly. The rest gets landfilled. (We've gotten worse at this over time; in 1990, we managed to recycle 35 percent of total US aluminum waste.) As for our other infinitely recyclable packaging material, less than one-third of glass gets recycled in the United States, versus 90 percent in a number of European countries.

Even with such anemic numbers, being diligent on the home front about recycling paper, cardboard, metal, and glass is well worth the effort. Where we really face-plant with recycling is with our plastic, a material that simply hasn't been designed to be recycled; and in the case of composites, they're born contaminated from the recycler's point of view. There's also the problem of theoretical recyclability and actual recycling—a distinction the packaging and beverage industries have worked very hard to blur. Soda straws, plastic grocery bags, disposable eating utensils, yogurt containers, take-out food clamshells, films, and bags are all theoretically recyclable but in fact are almost never actually recycled. And the reality of that, the reality we need to see and not unsee, is nowhere more apparent than on the grocery shelf.

The Story Our Trash Tells

Jenna Jambeck has been curious about trash since those first youthful trips to her hometown dump. Captivated even then by what people threw away and why, the young Jenna sensed that trash told a story—about what people consider waste and what they value, what they want and what shames them. That fascination with the story of waste has animated her education and work ever since, leading her to undergraduate, master's, and doctoral degrees in environmental engineering.

What sorts of stories does she find in waste? One is the story of transformation. We think nothing of putting an unopened bag of chips in our backpack or purse, Jambeck says, but the moment the chips are gone, the thought of stuffing the bag back in our backpack or purse is abhorrent. We want to get rid of it as soon as possible, which is why chip bags are such a common item of litter. The bag hasn't actually changed, yet it has transformed from pleasure to pariah. There's some deep mélange of habit, custom, and marketing that, Jambeck says, demands a reboot of the disposable economy. She sees a plastic industry profiting from a system it designed to compel us to accept wasteful products just to keep up with the fast pace of life.

Another story that consumes Jambeck is one of social and economic justice: how the plague of waste-related pollution falls most heavily on communities of color, the poor, and the least politically connected. This injustice is not limited to the obvious harms of smokestacks and oil wells, she says, but also includes the disproportionate burdens of plastic packaging waste. Jambeck has studied dollar stores, often the only sorts of markets located in poor neighborhoods and food deserts, where tiny packages of common products, from detergent to toothpaste to cereal, fill the aisles. They may be cheap at the cash register, but they are the most costly per unit of product one can buy, and each purchase is proportionately more packaging than contents. Not only is this a poor value, but it imposes the greatest burden of plastic waste on those communities that have the least resources available to cope with it, Jambeck says. Meanwhile, the industries that create and profit from this waste bear no responsibility.

Jambeck's doctoral research and thesis were centered on landfills, to the surprise of no one who knew her. And where else could the Trash Genius have her ultimate meet-cute moment with the man she would marry? She and husband-to-be Matt met while working on a research project together at . . . a landfill. "Love amidst the leachate," she says, laughing, referencing something only true trash nerds would know: leachate, the liquid muck that oozes from waste as it decays inside landfills, must be contained by liners and barriers, lest it contaminate soil and groundwater.

They drew the line at a landfill wedding, though, choosing flashy over trashy for their vows: Las Vegas.

She shifted her research focus to plastic pollution in the early 2000s as public concern about the problem was just gathering steam following news reports of a massive concentration of plastic waste that came to be known as the Great Pacific Garbage Patch. There was concern, yes, Jambeck realized, but good information? Not so much. No one could say how much plastic waste was going into the world's rivers and oceans, where it came from, or what kind of plastic was most problematic—which meant there was no way to attack the problem with any reasonable chance of success.

A couple of things seemed clear to her: The garbage patch was just the most visible part of the problem, an aggregation of large debris caused by the giant gyre currents that spiral through the oceans. Less visible but more pernicious were the microplastics that seemed to be everywhere. Without good data, trying to stem the tide of plastic pollution would be like trying to cure a disease without a diagnosis. Without good data, the idea that there was only one big garbage patch of plastic took hold—as did the idea that it might somehow be cleaned up and that would solve the problem, when in fact the real problem was much bigger, more diffuse, and much harder to locate. It was not just on the surface but also in the water column, on the seafloor, in the sediment, and in the sea life. Even the deepest point in any ocean, the Mariana Trench, nearly 7 miles down from the surface of the Pacific, is littered with plastic debris, including countless plastic grocery bags. But this sort of anecdotal evidence was not enough. Jambeck made finding conclusive data on the plastic pollution crisis, then acting on it, her life's work.

She realized that a global problem needed a global solution—a crowd-sourced solution. In 2011, Jambeck teamed up with computer scientist Kyle Johnsen on a National Oceanic and Atmospheric Administration–sponsored project to create a smartphone app called Marine Debris Tracker (now just Debris Tracker, because it covers rivers, lakes, and dry-land litter, too). Environmental activists, surfers, boaters, and others all over the world began sending reports of plastic debris they found in the water and on organized beach and riverside cleanups to Jambeck, who used the reports to construct a global database. Three years later, she set out on a seventy-two-foot sailing vessel to survey ocean plastic pollution in the Atlantic. The all-female, international crew of fourteen scientists, environmentalists, artists, activists, and sailors named the mission the eXXpedition, the first of many such voyages to research ocean pollution and ecosystems, and to promote female leadership in engineering, the sciences, and on the environment.

The following year, Jambeck coauthored the breakthrough scientific study that rocked the world with her 8.8-million-ton estimate of yearly plastic pollution. That is the rough equivalent of one large dump truck filled with plastic tipping its load into the ocean every minute of every day of the year—a line of 525,000 dump trucks that, bumper to bumper, would span the entire continental United States.

The varieties of plastics involved, their abysmal recycling rates, and the fact that some of them float, while other types of plastic sink to the seafloor, wreaking havoc out of sight, was as horrifying as the sheer quantity.

Type of Plastic	Recycling Number	Sink or Float in Seawater	Common Uses	USA Recycling Rate
Polyethylene Terephthalate (PET)	1	Sink	Beverage bottles, textiles	18.4%
High Density Polyethylene (HDPE)	2	Float	Gallon jugs, detergent bottles	10.3%
Polyvinyl Chloride (PVC)	3	Sink	Piping, siding	Negligible
Low Density Polyethylene (LDPE)	4	Float	Retail bags, plastic film, wraps	6.2%
Polypropylene (PP)	5	Float	Bottle caps, yogurt containers, toys	0.9%
Polystyrene (PS)	6	Sink (PS foam floats)	Utensils, frames, foam packaging	1.3%
Others	7	Nylon sinks	Fishing nets (nylon)	22.6%

Source: Jenna Jambeck, US Senate Testimony

Jambeck's subsequent research on the cumulative amount of plastic waste and its constant annual growth suggests that we are on track to double ocean plastic pollution by the year 2025 to 17.6 million tons a year. Jambeck has another visual image to help us grasp this amount of zombie waste: imagine recruiting enough people to stand on every foot of ocean coastline in the world (all 221,000 miles of it) with ten plastic grocery bags, each filled with plastic trash, and having them hurl all that garbage into the sea.

These findings, quoted often in the media, in government reports, and by other researchers, made plastic pollution a global priority. They also instantly turned Jambeck into an influencer, and she was soon speaking, researching, and traveling the world as a National Geographic Society–funded explorer, and as the US State Department's plastic pollution expert and ambassador. The MacArthur genius grant—$800,000 over five years—gave her the financial wherewithal to expand her research further.

Much of her work is now focused on her CAP program, which is what brought her to Tifton and several dozen communities around the world. In Miami, for example, her data helped the city focus its plastic pollution strategy beyond beach cleanups to include inland sources of litter, particularly along the Miami River and Biscayne Bay, and the region's storm drainage system, where plastic pollution begins its journey toward the ocean. Other CAP recommendations for Miami suggest regulating nonrecyclable plastics—bags and wraps, for example—through possible bans or incentives for replacing them with less wasteful and more readily recyclable products.

Every community plan is different, based on the local culture, values, and needs. And this sensitivity to local custom has made many communities eager to work with the Trash Genius. Her résumé and reputation may be intimidating, but her quirky warmth and gentle persuasiveness have left behind a trail of friends and community-science collaborators all over, from the banks of the Ganges River to the sachet market stalls of the Philippines and up and down the length of the mighty Mississippi River, where the entire Jambeck family of four spent an entire spring break camping in a mobile home on the trail of trash.

For her testimony before the US Congress, Jambeck developed a

Source: Jenna Jambeck, US Senate testimony

flowchart of plastic pollution and broke it down into the five areas between origin and ocean, where the flood and growth of disposable plastic waste can be stemmed.

Acting at the end point of the waste is the most obvious and perhaps the easiest strategy: litter cleanup and capture. This is an individual and community challenge: organizing street, beach, and river litter cleanups, and deploying filters, screens, and other capture devices to snatch plastic debris out of rivers and storm drains before it gets away.

Moving up the waste chain, Jambeck calls for improving global waste management. This requires a more systemic contain-and-capture strategy at landfills and recycling centers and during trash collection, and could include creating a Ridwell-like approach to take in and recycle the problematic plastics that now elude conventional recycling systems. It also requires acknowledging that the "recycle everything" approach is not only a failure but also isn't the wisest choice for all materials. There are growing concerns that recycling releases toxins and microplastics into the environment (as part of the process involves grinding up plastics, which releases large amounts of plastic dust). Recycled plastics may have even more toxic chemicals in them than new plastics, because those substances can be concentrated through the recycling process. Some of the more toxic plastics may be better off safely tucked away in landfills or banned entirely, though more study is needed to figure out which ones. And some unrecyclable packaging—multiplastic mailers or tuna sold in plastic pouches, for instance—can have lower overall carbon footprints than their heavier recyclable counterparts (paper envelopes and metal tuna cans).

Beyond these complications, the cleanup and waste management strategies aren't about fixing the disposable economy. They merely attempt to treat the symptoms of ever-growing waste more effectively. Jambeck's top three strategies, however, are about treating the disease by strangling the disposable economy, and all of them are related to the plastics industry itself. The most obvious, and most difficult, is to reduce disposable plastic production. Right now, the world is going in the opposite direction, with plastic production being ramped up by the fossil fuel industry every year. In the past ten years, plastic has gone from using 6 percent of the world's annual oil supply to 12 percent. And as fossil fuel companies fret over losing market share to electric cars, their commitment to an ever-expanding plastic-powered disposable economy will continue to grow unless they are forced to rein it in.

The next best thing on Jambeck's hierarchy of strategies is also in the hands of the industry: reengineering plastics to make them more recyclable and more environmentally benign (which probably means emphasizing newer plant-based plastics and killing those that are fossil fuel derived). We are nowhere near that point, either.

The third industry-focused strategy—shifting to reusability—can power all the others. It is the only one that can be driven by consumer choice, lowering demand and thereby forcing the industries profiting from the disposable economy to do what they refuse to do on their own: slow the tide of harmful plastic waste. All it takes to jump-start this shift is for individuals, households, school districts, small businesses, college campuses, and communities to embrace reusables and reject disposable containers and packaging whenever possible. Jambeck sees movement in that direction, the embrace of reusable coffee mugs, bags, and straws, and the fact that some business owners are embracing it as well, and even give customers who bring in their own containers a discount.

And now there is a way to turbocharge this shift, one that is both novel and yet a return to the past. It has a wonky name—extended producer responsibility, or EPR for short—and the idea is growing in popularity around the world. In the United States, this new method of taming the disposable economy has found its first test case in the communities of the Pine Tree State: Maine.

3

Message in a Bottle

Back in grade school, when Sarah Nichols wanted to persuade her mother to allow her to have a sleepover or to stay out late, she'd write a poem, stage an impromptu skit, strum a ukulele, or sing and dance—literally—to advance her cause. Sometimes the straight-A math student would create a form with checkboxes for her mom to fill out to make her case. She was the kid who gave multiple-choice tests to the parent.

Years later, she is still using her persuasive powers to get her way, though now Nichols's most beguiling song and dance is not about wheedling Mom—it's about fixing our broken recycling system. As the director of sustainability for the nonprofit Natural Resources Council of Maine, she set out to win passage of a revolutionary environmental law not by arguing about saving the climate or the oceans or the great outdoors, but by making it about saving taxpayers money—lots of it.

To say there was considerable skepticism about this bit of fiscal magic would be an understatement. Conventional wisdom stubbornly held that environmental good always costs us dearly, saddling us with higher prices or guilting us into giving up stuff we love and want—or both. The city councils and local businesspeople and state legislators Nichols approached were no different. Come on, Sarah, there's no such thing as a free lunch, they'd say. What's this really going to cost us?

Nichols would just smile in meeting after meeting and say nope, there would be no cost. Her plan was to end the massive invisible subsidy for waste that taxpayers have been saddled with for so long they couldn't see it, consider it, or realize it was there. Her plan was to *make* money for Maine's cities and towns.

"When my kids make a mess in the house, they have to clean it up," Nichols would tell legislators and council members and civic groups. "They have to take responsibility. But in the recycling universe, it's never worked that way. It's time that it did. We will shift the cost to where it should have been all along: to those who create the problem."

This simple analogy, this appeal to the fundamental rules of families that is also fundamental to capitalism—paying your own way instead of foisting costs off on others—started getting traction with those skeptical Mainers. Nichols worked on this for eight years, explaining that her idea wasn't a tax on businesses, as they would surely claim, but a long-overdue bill for picking up after their mess. She made her pitch, with plenty of data to back it up, at town council after town council, business by business, and during an endless number of rubber-chicken lunches and dinners with volunteer groups and civic organizations. Nichols's environmental organization is respected but small, so she recruited a statewide army of community volunteers to build support and spread the word about her recycling makeover at the local level. She calls this force her "Marges"—named for her first volunteer in an earlier environmental campaign. She defines a Marge as someone who's already an environmental advocate, but who needs some help on how to take action effectively. The Marges have become a force to be reckoned with in Maine, Nichols's not-so-secret weapon.

Few people get as excited by the topic of trash as Nichols, a single mother of two who spent three years after college as a Tahoe ski bum and bartender before earning her master's degree and returning to her native Maine. When she's riffing on waste stream percentages or ripping into industry arguments against producer responsibility laws, her features become mobile and animated, hands waving and gesturing, sometimes actually pulling her long blond hair in frustration or excitement before catching herself. She turns trash talk into performance art. "What can I say?" says Nichols. "I'm a trash super nerd."

When trade groups mounted opposition to the proposed law, arguing that, as she predicted, it would be a new tax, a job killer, and would raise prices for consumers, Nichols brought in experts from Canada and Europe, where such programs have existed for years without any such price hikes. And when one industry association argued that plastic wrap should be exempted from the law because climate change could be reduced by wrapping more food in plastic to keep it from spoiling and emitting methane, she literally laughed them out of the room. Maine towns were being bankrupted by recycling costs 67 percent higher than landfilling precisely because of problematic plastics such as bags and wraps, she said. Why should taxpayers be footing that bill?

Widespread enthusiasm for her idea finally came in the wake of blows delivered to recycling by National Sword and then the pandemic. Like other communities around the nation, towns in Maine either stopped or limited recycling. Even towns with strong zero-waste goals struggled, such as South Portland, population 26,498, which took a $100,000 hit to its budget. By 2021, Maine's town and city councils had become ardent supporters of Nichols's proposed new law. China's money had masked the fact that recycling wasn't working as promised. For decades, Maine had a legally mandated goal of achieving a 50 percent recycling rate. It has never gotten close. Now even its 35 percent rate was bankrupting the state.

Many small businesses and some larger ones in Maine joined Nichols's list of supporters, including popular state craft beer maker Allagash Brewing Company. The company's founder, Rob Tod, later testified to the legislature that he believed it was fair "to address the urgent issue of waste—in our forests, rivers, and oceans—by shifting the onus onto the companies that create it."

That was a turning point. The head of a popular, thriving business—the largest independent brewer in Maine—was saying that even he saw the sense in this. His company was part of the problem, he said. But no more. After all the meetings, setbacks, and yes, the occasional ukulele serenade, Nichols's vision for upending the disposable economy had gone from "not a chance in hell," as she was warned repeatedly eight years earlier, to a tantalizingly close possibility. Even the lobsterman who served as Maine's Republican leader became an ally.

Other businesses followed Allagash—not all, by any means, but the typical united front against regulation that the business community often erects was fractured. That's because this wasn't regulation. It was, Nichols argued, about restoring capitalism's core principle that both your profits and your costs are your own. We aren't changing the system, she said; we're just changing the billing address.

Nichols also knew enough to sweeten the pot by keeping it simple and giving small businesses a break. First, the law wouldn't change the way curbside collection and recycling centers work, but rather would preserve them by paying them to stay in business. Overseen by the state's Department of Environmental Protection, a new nonprofit stewardship organization would be created with industry participation—but not control—to collect fees from producers and sellers of disposable packaging and containers and then reimburse towns, cities, and counties for their recycling and disposal costs. Local governments would also receive grants funded by extra fees charged to producers whose packaging was deemed not readily recyclable and ended up in the landfill. Theoretical recycling would no longer cut it. And small businesses would be exempted from having to pay if they had less than $2 million in annual gross revenues or sold less than one ton of packaging per year to consumers in Maine.

Even so, trade groups and other opponents fought back with a rival measure, full of exceptions, carve-outs, and a "recycling rate" based on sales of packaging labeled as "recyclable," regardless of whether or not it actually got recycled. The rival bill created a stewardship organization run by the packaging industry itself. Characterized by Nichols as allowing "the fox to guard the henhouse" in her testimony to legislators, the rival measure failed in a straight-up vote by the state legislature.

Legislative sponsors from both sides of the aisle began lining up to support Nichols's vision of extended producer responsibility and its simple imperative that businesses must accept responsibility not only for the creation of their disposable plastic products but for their death as well. An idea once considered fringe now seemed not just obvious but also a matter of basic common sense. Several of these legislators marveled: Why has it taken us sixty years to see it?

Without that shift, manufacturers of waste have no incentive to change, improve, or stem the tide of plastic pollution, Nichols argued. The principle of "polluters pay" would, in her view, become the engine driving all three of Jenna Jambeck's top strategies for fighting plastic pollution: reducing the supply, engineering better materials, and making reusability a more attractive business opportunity.

Maine—home to 3,500 miles of coastline and the largest continuous forestland east of the Mississippi, the vast wild place of the Northeast that inspired and enthralled Henry David Thoreau two centuries ago—is an ideal test case for Nichols's plan. Mainers' pride in the green beauty of a state known both for its balmy summers and frigid winters crosses political lines. Nichols thinks the nature of Maine's legislature also made it fertile ground for an idea that failed in other states, such as California and New York. In Maine, lawmakers are part-timers. Most have day jobs and few are professional politicians. The power of lobbyists and political action committees and threats to support other candidates just doesn't have the same power over legislators who don't actually need the job, but do it as public service.

Nichols's bill was introduced at the statehouse with an overlong yet irresistible name: An Act to Support and Improve Municipal Recycling Programs and Save Taxpayer Money. It passed in 2021, the first law in the nation to hold producers responsible for the costs and pollution of the disposable economy. Other states have followed suit. It will no longer be enough to slap on a label that simply says "recyclable," a misleading term that promises nothing and, in the real world, actually means "recyclable in theory but rarely in practice." In Sarah Nichols's new world, "recyclable" means "will be recycled." Her name wasn't on the bill—the legislative sponsors are the ones who officially authored it and the state's governor signed it into law—and officially she's just a witness and an environmental advocate. But legislators acknowledged Nichols as the driving force, the one who had huddled in the hallways outside the chambers, advising, strategizing, lining up support, sometimes in the late hours of extended sessions. Nothing like this is ever the work of only one person, the legislators said, but it definitely would not have happened without one person.

There was a spate of news coverage the next day, including a story in *The New York Times* about Maine launching a new era of recycling. And then the battle began all over again.

A law is nothing without implementation—the fine print, the procedures, the execution of ideas in the real world. Creating the rules, forming the stewardship organization, staving off all the exceptions and loopholes that the beverage and packaging industries immediately started demanding will take years, and the first payment from the disposable-waste makers isn't due until 2026. Nichols has her work defending the new law cut out for her, although she has a bigger agenda.

This was just a start for her real goal, the one she put on the back burner until extended producer responsibility was in place. Nichols has shown us that one person—even one who hails from a small town and a small environmental group in a small state (ranked forty-fourth in population)—can transform national policy with her vision and persistence. She intends to do it again. She wants to use this shift in the disposable economy's balance sheet not just to fix recycling but to replace it with the one industry strategy Jenna Jambeck suggested in her congressional testimony that can be driven from outside.

Nichols envisions Maine as the first reusable state.

The Evolution of the Bottle—How Did We Get Here?

Some Americans can still remember, though most are now too young to have seen it themselves, that sustainable, reusable, and environmentally benign containers and packaging were the only game in town before disposable plastics arrived on the scene.

Our prosperity, progress, quality of life, and the country's standing in the world did not depend on shifting to single-use disposable packaging.

But that's what happened, and the story of why it did begins with the far more sustainable glass bottle, our first mass-produced packaging.

The glass bottle has been around for 3,500 years, though the art of making it was lost for five hundred years or so after the fall of ancient Rome. Rome's glassblowing techniques, copied from ancient Syria, finally were rediscovered in Europe around the year 1000. Bottle making first came to North America early in the 1600s, when a glass-melting furnace was built in the first permanent English settlement on the continent, Jamestown, Virginia. By the early 1800s America had eight glassworks— not a great number for a whole country. Bottles remained relatively scarce and valuable, handmade vessels prized not just as durable containers for goods but as goods themselves to be conserved and traded.

Beverages, like other goods, were shipped in bulk then, not in individual portions, as is common today. Brewers, wine sellers, pharmacists, and other merchants would either fill bottles brought in by customers or charge a higher price to supply a bottle—to be refunded with the bottle's return for refill. This functioned similarly to modern container deposits, but more effectively because a penny or two had far greater worth then. Bottles were so valuable in this era that they would be listed in wills and inheritances, carefully inventoried when the estates of the wealthy changed hands.

In the late 1800s, as the process of bottle-making became more of an assembly line using molds instead of freehand glassblowing, then grew increasingly automated, glass bottles became more plentiful, yet they remained valuable because demand grew even faster. A thriving trade in used bottles flourished in most cities to fill the void. Eventually, branded bottles were introduced, with inscriptions such as "This Bottle Not to Be Sold," and "This Bottle to Be Washed and Returned." Not everyone followed instructions, though, and need gave rise to a new career—bottle detective. Cadres of sleuths were hired to hunt down black marketers and home hoarders who trafficked in stolen bottles. As had long been the case, the contents of the bottle belonged to the customer, but the seller owned and was responsible for the bottle both before and after purchase. The reusability kept the price down for consumers because it lowered costs for the seller. And, bottom line, the waste of something useful was considered abhorrent, if not sinful.

A new beverage, meanwhile, burst on the scene to ramp up glass bottle demand even more—one of the first popular sodas, brought to New York City by Eastern European Jewish immigrants. The sparkling drink originated in the German town of Niederselters, where the popular carbonated beverage was called Selters Water, though the Yiddish variant stuck in the US: seltzer. By the early 1900s, elaborate soda fountains were popping up all over the country to offer seltzers plain and flavored, including a concoction created in New York called the egg cream (usually made with chocolate syrup, milk, and seltzer but, oddly, no egg). Soda fountains grew even more popular once Prohibition closed down legal bars and beer halls.

Around the same time, a seltzer bottle or two, then called "siphons," became common in many households. They looked a bit like fancy glass fire extinguishers with silver triggers and spouts, and the pressure of carbonation caused the fizzy water to shoot out. A flourishing industry emerged in which customers would bring in empty siphons and take full ones home, pressurized and ready to go. Delivery was also an option, just like the common milk deliveries of the era, exchanging empties for freshly filled. These siphons were multi-serve, not single-serve, and like ordinary bottles, they generally belonged to the seller, not the customer.

In time, branded bottles with flavored seltzer appeared, including an upstart new drink called Coca-Cola. Coke in its original iteration was a sustainable, low-waste product sold as a concentrated syrup. It was marketed as a health tonic—its original formula included a small amount of "medicinal" cocaine, long since removed—to be mixed with seltzer, which was also believed to have health-enhancing and antiseptic properties. The concentrate was sold mostly to soda fountains, though there were home customers, too. Some of the more successful soda fountains were rewarded by the company with fancy Coke-branded porcelain urns to dispense the syrup, one ounce of Coke for every five ounces of seltzer, plus crushed ice.

The advent of seltzer in branded bottles inspired the next phase in

Coke's evolution: the premixed single-serve bottle. Bottling Coke this way was a masterstroke for profits because it was a way of making more money for less product. Instead of selling 100 percent Coke syrup, bottled Coke was 90 percent carbonated water.

Ironically, the attractions of this single-serve innovation were lost on the president of Coca-Cola at the time, Asa Candler, who had bought the rights to the beverage from its inventor. Candler believed the future of the company lay with soda fountain sales. Convinced no one in their right mind would want to buy little 6.5-ounce single-serve bottles of Coke, Candler sold the bottling rights in 1899 to a pair of Tennessee lawyers for a dollar. He imposed only one condition: the price for a bottle of Coca-Cola had to remain a nickel in perpetuity—the same price soda fountains charged at the time, which made Coke a pretty good bargain even then.

Had Candler been right and the little bottles of Coke flopped, the world might be a very different place right now, and we might not be eating a credit card's worth of plastic each week. But the single-serve convenience you could drink anywhere captured customers' imaginations, and the new bottlers were ingenius at marketing it. Their advertising imagery was bold, colorful, and suggestive, showing people playing sports or working outdoors, then quenching their thirst with Coke. The catchy slogan that went with it: "The Pause That Refreshes." By the late 1920s, bottled Coke—with what would become Coca-Cola's iconic fluted green-glass bottles, bottled not by the company but by a contractor—was outselling the soda fountain variety.

But the company made up for Candler's decision by deploying nearly a half million branded vending machines nationwide to sell massive numbers of those little bottles, so many of those nickels came to the Coca-Cola mothership anyway. The brand had pioneered an essential building block of a more wasteful economy of the future. Consumers had been cajoled into forsaking an undiluted product sold in highly concentrated, multiserving form, or consumed as a treat while socializing at a soda fountain, in favor of a version that was mostly water in a single-serving bottle sold as a solitary impulse buy. With the right marketing, the soda maker learned, perceived convenience could trump value, and customers would willingly pay for flavored water over and over, rather than a single economical bulk purchase that could be mixed with soda water at home far more cheaply. The rest of the business world took note, and many other industries in the years to come adopted this approach: notably manufacturers of shampoo (originally a powder that the purchaser mixed with water), liquid soap (supplanting the once dominant bar soap with 71 percent of the 2023 market, and, of course, plastic bottles instead of a paper wrap), and laundry detergent and dish soap (both once sold as concentrated bars or flakes). And buying flavored sparkling water in plastic bottles has become normal, a habit we are asked to accept as the way things are supposed to be, perhaps even *must* be—even though all these products started out in waterless, concentrated forms, and the way we do it now gives consumers less value, while adding tremendous amounts of waste, shipping, packaging, and pollution.

However, throughout this transition from sustainable, low-waste product to diluted single-serve beverages, the glass bottles still belonged to the sellers, not the consumer. Coca-Cola, its competitors, and the brewing industry were not yet ready to try to wriggle out of long-standing responsibility for the fate of their packaging—taking the bottles back, washing them, and reusing them twenty to fifty times, after which they could be ground up and recycled. Coca-Cola and its contract bottlers and shippers were deeply invested in a nationwide network of regional centers to handle this efficiently, with automated plants that had the fascinating clockwork precision of mechanical pinsetters at the bowling alley.

In 1948, Coca-Cola moved to make this strategy even more sustainable by imposing a two-cent deposit on each bottle—which was pretty stiff, given that the drink itself still only cost five cents for those little bottles. But sales continued unabated, and the results demonstrated the power of deposits to prevent waste. Ninety-six percent of bottles were returned so that consumers could get their two cents back—which more often than not went toward the purchase of more beverages. So even though it was single-serve and customers were getting less for more, this remained a circular, low-waste business when it came to packaging. That lasted right up to the early 1970s—the last time plastic waste in America's trash cans would be an oddity rather than the norm. As for whatever benefits the subsequent emergence of the disposable plastic economy would bring us, it wasn't reflected in our pocketbooks. The end of the reusable age, coincidentally, also marked the point when the income and buying power of wage earners in America, adjusted for inflation, reached its highest mark before or since.

The three-century-long consensus that sellers, not consumers and taxpayers, were responsible for the fate of their packaging began slowly unraveling in the 1950s. That's when the major national-brand beer brewers, followed by some soda makers, stopped taking back glass bottles—the beginning of the "no deposit, no return" era. At the same time, both industries also started offering their products in disposable cans, and in 1970 the first plastic soda bottles arrived on the scene.

One by one, state by state, the bottle washing and filling plants closed down. One of the last to go was the Kleis Beverage Company of Ann Arbor, which had been bottling sodas in reusable glass since 1907 and which, in the 1950s, washed and filled more than 8 million returnable bottles of Coke a year. By the 1970s the plant, still run by the Kleis family, was turning out half its colas in reusable glass, and half in no-return containers. Owner Harold Kleis was not pleased by the transition and seemed to think consumers were crazy to embrace it. He told *The Ann Arbor News* that he'd provide customers whatever they wanted, but they were paying a penalty of seventy-two cents a case for choosing sodas in throwaway bottles. Even so, the deposit system for reusable glass quickly dwindled. Profits soared as the beverage and bottling industries abruptly shifted the cost of cleaning up after their no-longer-reusable packaging from themselves to . . . nobody.

The results were predictable: cans and bottles started to fill trash bins and litter the landscape, beaches, and waterways at an alarming rate. Coca-Cola, meanwhile, emerged as the number one source of disposable plastic bottles in the world.

Civic leaders and legislators proposed the obvious solution: mandatory deposits and outright bans on nonreturnable glass bottles. At one point in the 1960s, twenty-three states and the US Congress had proposed laws to force industry to take back responsibility and stop unnecessary waste. Through the mid-1970s, there were more than a thousand failed attempts to ban or tax disposable packaging in order to revive the network of bottle washing and filling plants and stem the tide of waste.

The beverage and packaging industries fought back with lobbyists, pleas, and their most devastating weapon . . . the "Crying Indian."

This is one of the most famous advertising campaigns in history, in print and on television. It debuted on the second Earth Day, April 22, 1971, part of the Keep America Beautiful organization's "public service" campaign, and its message had a powerful and long-lasting effect on millions of viewers.

It begins with dramatic bass drum rhythms and a man in what is supposed to be traditional Native American garb, wearing leather buckskins and a feather in his long, braided hair, paddling a canoe down a lovely woodland river. The drums are joined by haunting low notes from a bass violin, then a full orchestra, the music swelling as the camera shifts focus from the paddler to the water, revealing that the canoe is passing through increasing amounts of floating trash and debris. The camera pans back, showing that the canoe has moved onto a stagnant lake, the shores crammed with factories and smokestacks spewing noxious fumes. The music reaches a frantic brass-heavy crescendo, then shifts to a quiet, plaintive flute as the man lands the canoe and strides across a rocky, litter-strewn beach covered with bottles, papers, and food wrappers. This is when the narration begins. The deep voice of a well-known actor of the era, William Conrad, intones, "Some people have a deep abiding respect for the natural beauty that was once this country . . ."

As the voice pauses, we see the canoeist is now at the edge of a lakeshore highway choked with traffic, where someone in a passing car carelessly drops a bulging fast-food bag out the window. As it falls and spills garbage at the man's moccasined feet, the narrator continues, "And some people don't."

The camera zooms in on the man's face, and the music swells to a final crescendo as he turns to look directly into the camera, a single tear running down his seamed cheek.

"People start pollution," concludes the narrator. "People can stop it."

For many Americans at the time, and for many years after, this was the most powerful and moving environmental call to arms they had ever seen. Even all these years later, the comments on YouTube discussing the ad forever known as the "Crying Indian" still reflect the near reverence many feel for this message:

"It shaped my appreciation and desire to protect the environment."

"I remember seeing this in the 70's when I was about 3 or 4 years old. I still am in awe of this commercial. It breaks my heart still to this day 50 years later."

"This commercial always makes me cry. Even if I wasn't of Native heritage, it would still impact me. . . . He's so sad to see his world polluted and treated like nothing more than property rather than a wonderful living thing . . . but it feels like he knows that he can't change things. Not anymore."

"One of the most powerful commercials I've seen. Drives the message home."

But there is more to this moving one-minute slice of television history than meets the eye. Most who revere this legendary advertisement don't know that the litter seen in this ad was carefully curated to avoid any brand names that could implicate the beverage and packaging companies that, it turns out, paid for this commercial. Or that its official creator, Keep America Beautiful, was an industry front group whose mission was to oppose bottle bills, deposit laws, and any legislative or public sentiment that sought to blame litter and plastic pollution on the makers of wasteful products and packaging. Or that the "Crying Indian" character was actually played by the late Italian American actor Espera Oscar de Corti, who used the stage name Iron Eyes Cody, appeared in more than two hundred films and TV shows, mostly Westerns, and often wore his (not authentic) costume in public whether he was working or not.

To head off laws forcing producers to resume paying for the fate of their packaging, these industries spent millions on such advertisements blaming citizens, not companies, for pollution. *People start pollution. People can stop it.* The idea was to focus Americans on volunteer litter cleanups and volunteer recycling programs, marshaling schools and scout troops and civic groups to round up trash in parks, beaches, and playgrounds, rather than blame the shift from reusable to disposable bottles and plastics. The advertising geniuses who worked for Keep America Beautiful came up with the term "litterbug" and the slogan "Every litter bit hurts" to lay blame for pollution and plastic waste on consumers rather than containers. Coca-Cola also launched an ad campaign pushing the phrase "Bend a little," complete with free branded litterbags it gave out nationwide to encourage people to spontaneously bend over and pick up litter. Cartoonist Charles M. Schulz, creator of *Peanuts*, even drew a poster aimed at schoolkids in 1972 featuring the character Woodstock carrying a sign saying BEND A LITTLE, PICK UP A LOT, which the US Department of the Interior published and posted nationwide.

And then came the campaign's crowning achievement, the "Crying Indian" ad and a host of variations and related imagery, which fulfilled Keep America Beautiful's mission to shift blame and financial responsibility for packaging waste from industry to taxpayers. They framed this as free enterprise at work, but they had, in effect, socialized their waste and kept all the profits for themselves. At the same time, the "Crying Indian" ad's message was that those same taxpayers were to blame for the problem, and that individual action, not producer responsibility laws, provided the only solution. The "Crying Indian," in short, was the single most effective piece of greenwashing in history. (In 2023, the ad, long criticized by Indigenous

Americans for stereotyping and cultural appropriation, was retired and its rights transferred to the National Congress of American Indians.)

The terms of the conversation about waste and its solutions have changed little since 1971. Most of the bottle bills proposed back then failed. One that managed to get passed, requiring reusable bottles for beer in Vermont, was allowed to sunset after three years. Even in the ten states where bottle bill deposits eventually were adopted, the onus for collection and recycling materials was then and is still now on the taxpayers. The message that it's people, not producers, who are responsible became the norm, even when it comes to climate disruption and ocean plastic pollution. The "Crying Indian" reframed multiple environmental issues and overturned centuries of consensus, norms, and history.

That reframing is what Sarah Nichols finally succeeded in challenging, and Maine's new law started an avalanche. Five other states have adopted comparable regulations: Oregon, Washington, Colorado, New Jersey, and most recently California, where extended producer responsibility had stalled in the statehouse for years. A total of seventeen other states have been debating EPRs for plastic containers and packaging as well.

Cities are taking a different approach, with Seattle leading the way with a push for reuse. Partnering with the nonprofit Upstream and the reusable beverage company r.Cup, Seattle tested a reusable-cup program at the city zoo, with container drop-off stations for washing and reuse. Now the city is expanding into concert venues, with sports stadiums—and their massive amounts of container waste—up next. Once the community gets used to the system of reuse versus disposables at these "contained" gated venues, where collection is easy, Seattle plans to roll out reusable programs for street businesses—coffee shops, restaurants, movie theaters, and malls. Seattle is banking on the belief that the solution to an ineffective recycling system is an effective reuse system.

Nichols says we need both, but her top priority is defending her law, which will remain vulnerable until it finally goes into effect between 2025 and 2026. Part of that, she says, is cracking down on deceptive and confusing labeling that makes every plastic container and package seem recyclable to consumers, when the opposite is the case.

"The labeling of plastic packaging can be compared to the lawlessness of the Wild West," says Nichols. "Most of it is adorned with images of the earth, flowing blue streams, freshwater springs, green leaves, trees; words like 'eco-friendly' and 'sustainable'; and the most confusing of all—the recycling symbol."

Those chasing arrows are put on everything, she complains, even when most communities cannot recycle most of the types of plastic that support that symbol. Nichols wants Maine to use its new law to fight deceptive labeling and hopes that other states hold the line as well. She argues, once again, from a position of common sense: only packaging that can actually be recycled should be promoted as recyclable. Anything else is consumer fraud.

It's the Wizard of Oz trick all over again, she says: "'Pay no attention to the man behind the (plastic) curtain . . .' But we *are* paying attention now."

Trash and Plastic: Where We've Been and What We Can Do

When my book *Garbology: Our Dirty Love Affair with Trash* was published in 2012, everyone thought they knew how much trash Americans throw away. The official Environmental Protection Agency figure—used by environmentalists, businesses, and policymakers—maintained that the average American rolled just over 4.3 pounds to the curb every day. About a third of that got recycled, the EPA said, with the rest going to landfills. The numbers were right out of the agency's exhaustive annual compendium, then known as *Municipal Solid Waste in the United States*, its "trash bible"—now expanded and rebranded with a more upbeat title: *Advancing Sustainable Materials Management: Facts and Figures.*

The problem is that the gold standard of garbage was (and remains) wildly wrong, leaving 140 million tons of refuse unaccounted for. In actuality, Americans were throwing out a bit more than seven pounds a day, sending nearly twice as much waste to landfills as the EPA let on. An obscure but far-more-accurate annual survey made jointly by Columbia University and the trade journal *BioCycle* did what the EPA's trash bible did not do and has never done: actually count our trash poundage using real-world data from the nation's landfills.

The EPA has relied largely on industry-provided data on product sales and estimates of how quickly those products wear out and get thrown away. This method was developed decades ago when there were eight times as many legal dumps, many more illegal ones, and little good data available. That picture has changed: there are now far fewer landfills, and most of them carefully weigh each incoming can, bottle, spoiled avocado, old couch, and crumpled candy wrapper—their pay-by-the-ton business model depends on it.

The Columbia-*BioCycle* surveys also revealed that Americans recycle or compost proportionately far less than the official stats suggest: not the third of our total trash estimated by the EPA—a milestone we were supposed to have surpassed a decade ago—but less than a quarter.

Nickolas Themelis, then the director of Columbia's Earth Engineering Center and now a professor emeritus, has long been one of the trash cognoscenti calling for reform of this "untenable situation." Among myriad problems created by the incorrect data, he frets, is the false impression that current waste-reduction strategies are working.

So now let's fast-forward to more recent times. Where do we stand now? If you think we've made progress, think again. We are more trashy than ever.

The EPA's trash bible, which as of the end of 2023 was still using 2018 pre-pandemic data, puts America's annual municipal solid waste (MSW)—our total trash—at 292.4 million tons.

Less than a quarter of that was recycled: 23.6 percent. Another 8.5 percent was composted. And the EPA claims that about half our waste—146.2 million tons—was landfilled that year. (The remainder—about 18 percent—was reported to have gone to waste-to-energy plants and other uses for food waste, such as biofuels and animal feed.)

EPA's Latest Trash Data Is Total Garbage

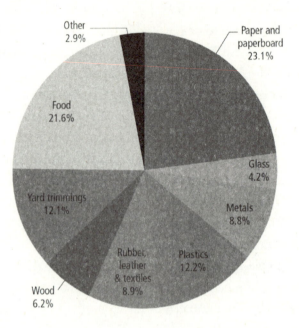

Other
2.9%

Paper and
paperboard
23.1%

Food
21.6%

Glass
4.2%

Yard trimmings
12.1%

Metals
8.8%

Rubber,
leather
& textiles
8.9%

Plastics
12.2%

Wood
6.2%

Source: EPA, "National Overview: Facts and Figures on Materials, Wastes and Recycling," December 2022

Based on all this, the EPA reported that the average American throws away 4.9 pounds of trash each and every day—up by a bit more than a half pound a day from when I wrote *Garbology*.

None of these numbers are true. The EPA continues to use the same flawed methodology and put out incorrect numbers that make it appear we are far less wasteful than we truly are, and that we recycle and compost at higher rates than we do.

The annual *BioCycle* analysis of trash ended in 2013, but Columbia still collects some of that data, including how much measured waste enters US municipal landfills. Using the most recent data available, for 2017, Columbia's Earth Engineering Center reports that Americans actually throw away 200 million tons more than the EPA estimates. The real numbers put Americans' trashiness at 8.2 pounds a day for every man, woman, and child.

The actual percentages for recycling are also much more anemic than the EPA's calculations. We recycled only 14 percent of our waste, while landfilling more than 70 percent.

Even the EPA acknowledges the lack of true revelation in its trash bible. A separate department at the agency that calculates the climate impact of our waste uses a much higher landfilling number that splits the difference between the EPA's official stats and Columbia's. That calculation puts Americans' daily trash production at 6.5 pounds.

Any way you cut it, the numbers are not an accurate reflection of reality, and we remain one of the trashiest societies on earth. And the single largest category—28 percent of what we throw away—is containers and packaging, the disposable economy, which is where most of our plastic garbage resides.

WHAT YOU CAN DO RIGHT NOW ABOUT THE DISPOSABLE ECONOMY

Refuse plastics.
- Don't accept plastic bags when you shop anywhere. That includes the little produce bags at the grocery store. Supermarkets with bakeries make fresh breads that you can bring home without plastic, too.
- Just say no to plastic utensils, plastic takeout containers, plastic-foam anything. If your favorite takeout places insist on using nonrecyclable containers, ask them to switch to compostables and paper sacks—and if they won't, find places that will.
- Avoid all single-use plastic beverage bottles. If you must have single-use sodas, buy them in cans or glass, or consider a soda-making device such as SodaStream. *Don't* buy plastic bottles of water. Chill your tap water and drink that—in most places, the quality is as good as or better than bottled water. If you're worried about water quality, get a filtration pitcher or faucet attachment.
- Buy alternative forms of common products to avoid plastic containers: toothpaste tablets instead of toothpaste in a tube, bamboo

toothbrushes instead of plastic, shampoo and conditioner bars instead of disposable plastic, laundry detergent bars or sheets. Or buy these liquid products in bulk and bring your own bottles to a refill store.

- For food storage, avoid plastic wrap, plastic bags, single-use plastic containers, and conventional wax paper, which is unrecyclable and is often made from fossil fuel–derived paraffin wax. If you absolutely must have a wrap, there are some made from beeswax, or you can use parchment paper or go with aluminum foil, which can be pinched to make an airtight seal, then cleaned and reused or recycled. Just make sure you save your used foil until you have enough to compress into something close to a baseball's size and shape before tossing it into the recycling bin. Too small and it might not be sorted correctly.

Embrace reusables.

This is the biggie: *Reuse* is the opposite of the disposable economy. Bring your own reusable bags whenever and wherever you shop—not just the grocery store, but when you shop for anything: shoes, clothes, gifts— anywhere that puts stuff in a plastic bag.

- Use reusable cups for coffee or any other single-serve beverage you buy.
- Bring reusable containers to purchase nonpackaged foods: coffee, loose-leaf tea, deli meats and cheeses, baked goods, takeout, and restaurant meal leftovers, just to name a few. Who needs a doggie bag when you have your doggie container?
- If you're refusing plastic wraps, bags, and other single-use plastics for food storage at home, there are zero-waste options. Reusable containers of glass, ceramics, and stainless steel are easy to clean and can stack in the fridge. Reusing pickle, spaghetti sauce, mayonnaise, and peanut butter jars are great ways to do it on the cheap. Keeping food fresh in the fridge can be as simple as putting it in a glass bowl and setting a dinner plate on top—no plastic wrap needed.
- Pro tip (because it's easy to forget): keep a zero-waste kit in your car, bike bag, backpack, or whatever you use when you go out so you are

ready with reusable alternatives to common disposables: bags, cups, containers. For tips on assembling a zero-waste kit at no cost, see this post at author Anne-Marie Bonneau's *Zero-Waste Chef* blog: zerowastechef.com/2018/01/28/a-zero-waste-kit-that-costs-zero -dollars. And check out her other tips on cutting waste and living plastic-free, too.

Recycle the packaging you do buy.

- Paper, cardboard, cans, and select plastics are recyclable, but the rules vary from one community to the next. Check with your local municipal waste agency or private trash service to see what you can put in your recycling bin or, if your community has no curbside recycling, to find drop-off locations.

- If you live in one of the ten container deposit states, bring those items to a redemption center (or, in some states, the store where you bought them) and get your money back. If you toss them into the recycling bin, the likelihood that they'll actually be recycled drops drastically, plus the state will keep your deposit money. Here's how to find out where to return deposit containers in your state:

  - California: The state recently expanded deposit rules to cover wine and distilled spirit bottles. Basically all beverage containers now require a deposit of five or ten cents, so it pays to get back that cash. There are official redemption centers in some areas, but for communities that lack them, stores that sell single-serve beverage containers have to accept the empties and return deposits. To find your nearest redemption location by zip code, use the search tool at calrecycle.ca.gov.

  - Connecticut: For the nearest deposit redemption center, go to portal.ct.gov/DEEP/Reduce-Reuse-Recycle/Bottles/Connecticut -Redemption-Centers. All beverage containers in Connecticut have deposits, whether cans, bottles, jars, or cartons made of plastic, glass, or metal.

- Hawaii: Redemption center locations can be found at health
  .hawaii.gov/hi5/redemption-centers. The Aloha State has deposits
  on mixed spirits (coolers), beers, malts, and all types of
  nonalcoholic drinks except dairy.
- Iowa: All beverage containers, including hard liquors, beers,
  wines, soft drinks, and water, have deposits, no exceptions.
  Your nearest redemption center is here: programs.iowadnr.gov
  /solidwaste/bottlebill/map.
- Maine: One of the first bottle deposit states in the modern era,
  Maine requires deposits on all beverages except dairy products
  and unprocessed ciders. The state does not maintain a list of
  active redemption centers, so you'll have to rely on a web search for
  nearby ones.
- Massachusetts: With one of the lowest return rates of the bottle
  states (38 percent), Massachusetts needs your help in ramping up
  its bottle and container deposit program. Find out where to take
  your empties here: mass.gov/info-details/find-a-bottle-can
  -redemption-center.
- Michigan: The deposit law here covers all containers made of any
  material for any beverage except wine and hard liquor (unless it's
  inside a cooler). Michigan also makes returning easy: you can
  bring containers to grocery stores, gas stations, supermarkets, or
  convenience stores—which, before disposable plastic beverage bottles replaced reusable glass, was the nationwide norm.
- New York: Deposits on any type of container for wine
  coolers, water, soft drinks, carbonated drinks, beers, and malts are
  required and refundable. New York doesn't provide a redemption
  center list, but a web search for nearby New York State bottle
  redemption centers should do the trick.
- Oregon: The country's first bottle deposit law in the modern era—
  in place since 1971—is also the most successful, with an
  81 percent return rate. All beverage containers and types are
  covered except wines, hard liquors, milk (both dairy and

plant-based), and infant formula. Find your "BottleDrop" here: bottledrop.com/locations/.

- Vermont: All beverage containers in the Green Mountain State are covered except wine, unless it's in the form of a wine cooler. The official redemption center list is here: dec.vermont.gov/sites/dec /files/wmp/SolidWaste/Documents/RedempCenters.pdf.

LONGER-TERM SOLUTIONS

- Volunteer for beach, park, and river cleanups.
- Use the Debris Tracker app to help solve the plastic pollution crisis.
- Form your own neighborhood Owen's List to get hard-to-recycle items to places that will recycle them.
- Support proposals in your state and community to ban disposable plastics, to adopt extended producer responsibility laws, and to adopt container-deposit laws. Also vote for politicians who favor such policies.
- If your community or local school district makes little or no effort to recycle, reuse, or reduce packaging waste or to favor sustainable snacks (for instance, in school cafeterias or at local sports venues), lobby your elected officials to enter the twenty-first century and start dealing with the plastic waste problem. There's no excuse for a school or a town not to have an active recycling program. Sometimes all it takes to get the ball rolling is to bring it up during the public comment portion of a town council or school board meeting.

10 TIPS FROM SARAH NICHOLS ON HOW TO BE A MARGE

1. **Believe in yourself.** You have more power to effect change than you might think. Big systemic problems can be overwhelming, but local policies lead to state policies, and those can lead to federal policies. And that can impact the entire world.

2. **Trust that your voice matters.** If you've never done it, it can be intimidating to talk to authority figures and elected representatives. But they are people just like you, and they won't always be knowledgeable on every issue. You can serve as their trusted source of information on issues that you care about.

3. **Brace yourself and be nimble.** Policy change can be a roller coaster. It might take years to reach your goal, and you might need to change strategy along the way. Know where you must stand firm and where you're willing to compromise.

4. **Do your homework.** Learn until you feel like you can answer any questions with ease, or at least know where to find the answers. This will give you confidence and clout as you work to get others on board. Passion can be infectious.

5. **Don't reinvent the wheel.** You're likely to find that there are other people pushing for the same change. Get in touch with them and share information and resources with one another.

6. **Find your people.** Identify a few people who care about your goals as much as you do, then work together closely with your champion on the town council, the state legislature, or the local advocacy organization who shares your concerns and your mission. You will need each other to get things done.

7. **Meet people where they are.** All voters and decision-makers come with their own unique perspectives. Listen to them and find out where you have shared values. The reasons why you might both support the same policy could be different, but the outcome can be the same.

8. **Remember the 20-60-20 rule.** There will always be 20 percent of people with you and 20 percent against you no matter

what. You want to focus on and appeal to that 60 percent in the middle.

9. **Involve young people—it's their future.** They are smarter than you think. Community leaders love hearing from them, and in the process those young people learn invaluable skills and confidence.

10. **Celebrate, then steward the policy.** Passing a policy can be thrilling and worthy of celebration, but just like a new baby being born, you must then raise it and make sure it's a healthy and strong member of society. See it through!

PART II

Power Hungry

Do not be vulnerable to despair. We are going to do this. And if you doubt that we as human beings have the will to act, please always remember that the will to act is itself a renewable resource.

—Al Gore, TED Countdown Summit, July 2023

4

Ring of Fire

We all live in a house on fire.
—TENNESSEE WILLIAMS

The cuisine was incredible, the views spectacular, the customers happy, but behind the scenes at one of Pittsburgh's top restaurants, the kitchen was just like most eateries': a toxic sweatshop. Rows of gas stoves, ovens, and broilers cooked not just the food but the whole room—and everyone in it. Meanwhile, they spewed harmful emissions at levels that would be considered illegal outdoors. But that didn't help Chef Christopher Galarza as he worked a sizzling flat-top griddle with a roaring broiler at his back. Air pollution laws stop at the doorway. Indoors—not in all countries, but definitely in the United States—is the Wild West of air quality.

One busy night, the heat and fumes inside the restaurant kitchen became so intense that Galarza and his crew members resorted to guzzling bottle after bottle of water to stay hydrated. They were all drenched with sweat. As the dinner service wore on, they began taking turns racing to the bathroom to heave. On a whim, Galarza stuck a meat thermometer in his tunic pocket. Just to see.

A talented young chef just starting to make a name for himself in the Pittsburgh foodie scene, Galarza had worked hard to land this gig. He loved being a chef. The son of a Brazilian immigrant mom, Galarza grew up in New Jersey, where he weathered several bouts of homelessness before the family moved to a more stable life in Florida. Galarza found a job as a busboy there and though he was supposed to be working the front of the house, he constantly hung out in the kitchen. He'd ask the cooks about the work and spent his downtime reading the restaurant copy of the Culinary Institute of America's textbook cover to cover, until they finally gave him a shot at cooking. Three years later he left for culinary school at the Art Institute of Pittsburgh, and after graduating he began working his way up in the world of fine dining. He knew what every ambitious young chef knows: enduring the overheated crucible of a busy, high-end kitchen was a test—part assessment, part hazing, part Darwinian selection, like sleepless medical residents always on call or new cops stuck working the crappiest, most backbreaking shifts. It was tradition. And he had survived, and thrived, working in several legendary master chefs' kitchens. Now he had a dream job—except for that thermometer burning in his tunic pocket.

After a few minutes, Galarza checked the temperature: 135 degrees Fahrenheit.

Ambient temperatures that high can be deadly. Water that hot can scald a child's skin in four seconds. Five degrees higher, and vinyl plastic starts to warp. All that heat was waste—in fact, 70 percent or more of the energy used by a typical gas stove is wasted cooking everything in the room other than the food. Galarza felt awful. He soldiered on, though the seeds of discontent had taken root. There had to be a better way, he thought. He just couldn't imagine what that might be. Not yet.

He eventually left the restaurant for a stint as a catering chef at Carnegie Mellon University, making meals for the school president's events and fundraisers. That led to another higher-education job offer in 2015, one that would change the course of his career and life. He was tasked with creating and running an all-electric, climate-friendly, nonpolluting, zero-waste gourmet university cafeteria at Eden Hall, a four-hundred-acre campus dedicated to sustainable living at Pittsburgh's Chatham University.

Eden Hall is one of the most sustainable campuses in the world: its solar farm and geothermal energy generate more power than the campus uses, its buildings are LEED (Leadership in Energy and Environmental Design) certified, a nature-mimicking treatment system handles its waste, and organic food is supplied by its campus farm, greenhouses, trout farm, and experimental gardens; it even taps syrup from its own maple tree grove. The school wanted Galarza to be executive chef using those ingredients in a state-of-the-art sustainable kitchen—the first all-electric campus kitchen in America. Galarza, proclaiming the place "a chef's playground," jumped at the offer, even though like most every US chef at the time (and to this day), he could scarcely imagine cooking without fossil fuels and open flames.

But then he discovered the electric induction stove—cooking with magnetism. Not to be confused with those old-school electric stovetops with the metal coil burners that glow orange and that no self-respecting chef would touch, induction is another technology—billed as the most efficient way to cook food ever invented. Induction cooking does nothing hazardous or exotic; it has nothing to do with microwaves, nor does it heat through the process of electrical resistance, which is how original electric stoves work—the same process that starts fires when there's an electrical overload or a short circuit. Galarza thought it sounded too good to be true. He had heard of induction, of course—the technology was growing in popularity in Asia and Europe, where it was the mainstay at the world's top-rated restaurant, Copenhagen's Noma. In North America, Iron Chef Amanda Cohen had used induction at her Michelin-starred vegetarian restaurant, Dirt Candy, in New York for more than a decade, and Vancouver's king of Thai cuisine, Angus An, swore by his magnetic stoves. But they were outliers—their ideas about food were influential in the industry, but that didn't carry over to their choice of induction over fire. Galarza hadn't tried induction himself, and he just didn't get the allure. How could this cold ceramic surface with the digital controls replace cooking with fire and its primal nature? Any chef worth their salt could intuit the temperature and the sear by just the height of those blue flames. They'd adjust the heat with a twist of a good old analog knob that had no temperature settings on it, because a skilled chef needed none, relying instead on instinct, experience, and the feel of that smoky sizzle and heat rising from a pan.

But then Galarza went from skepticism to awe in less than a day. Induction heated a pan more quickly and precisely than the most high-end gas stove. Water boiled in half the time. Yet he could put his hand on an induction "burner" turned up high and not feel a thing. The kitchen stayed cool. Spatters didn't sizzle and smoke or get baked onto the cooking surface, because the stovetop stayed cool, too, which made cleaning a snap instead of the time-consuming breakdown and scrubbing with harsh chemicals that gas ranges require. This type of stove also reduced the risk of kitchen fires to next to nothing. No more racing to the bathroom to throw up from 135-degree temperatures.

What Is Induction, and How Does It Work?

Fascinated, Chef Galarza read up on the science of induction. Was there a limitation he wasn't seeing? A hidden drawback to this seeming miracle of new technology? Instead, he learned that there was nothing new about electromagnetic induction. It was discovered and tamed in 1831 by the English scientist Michael Faraday, who discovered a fundamental property of electrical current: it creates a magnetic field as it flows through a wire. Then he figured out the world-changing part: this process works in reverse. Magnetic fields can *induce* an electric current in wires without coming into physical contact with them. Faraday's law of induction would explain a key technological discovery of the modern age, as he had found a source of energy, light, and heat that didn't require burning something. Faraday used induction to create the first generator, and his invention lies at the heart of our grid to this day. In a generator, a turbine driven by steam, wind power, hydropower, or Faraday's original hand crank is used to rapidly spin a magnet at the center of a coil of wires. The continuous oscillation of the positive and negative poles on the magnet induces an electrical current to flow from the coil, transforming mechanical energy into electrical energy. That's what lights our lights, and everything else.

It's hard to imagine the stunning impact of something that now seems an ordinary fact of life, but the human world was forever changed by Faraday's discovery, which is why he is considered one of the greatest experimental physicists of all time. He had harnessed an elemental force. He gave us control of the lightning.

And the generator was just the start. Electric motors basically reverse the process, turning electricity into mechanical energy. Without induction, there are no electric cars, no power drills, no blenders or washing machines. There's another kind of induction that doesn't require spinning magnets to create electricity. Instead, it relies on the fact that our home electrical outlets deliver *alternating* current, which constantly oscillates all on its own between positive and negative, creating the same effect in a coil of wire as spinning magnets. This form of induction lets transformers change a dangerous high-voltage current into lower, safer voltages—big transformers atop utility poles, as well as tiny ones inside many devices, from home computers to model train sets. This is also how wireless phone charging works: the AC current flowing through a metal coil inside the charger projects an oscillating magnetic field upward, inducing a current inside a phone positioned above, which then flows into the phone battery and charges it.

Have you noticed that these chargers and your phone get quite warm in the process? In the charger, that heat is an unintended by-product of induction. In an induction stove, the heat is the *main* product.

Induction was first introduced as a means of cooking in the "Kitchen of the Future" display at the 1933 Chicago World's Fair. "Cold cooking" astonished onlookers by making a teakettle boil while sitting atop an ordinary book, which suffered no damage as the magnetic field passed through it harmlessly. It would take sixty years to turn this sideshow attraction into a reliable consumer product in which a powerful oscillating magnetic field could be created on induction "burners" beneath a sturdy ceramic glass stovetop. The magnetic field it projects focuses on the bottom of a pot or pan set down just like a phone on a charger. Once the current flows into the cookware, it has nowhere to go—no battery to charge, no blender to spin—so those electrons bounce around in the bottom of the pot, colliding with one another and creating powerful heat very quickly. The pot, not the stove, serves as the heating element. This explains why induction is so much faster and more efficient, using less energy to accomplish the same cooking, compared with other technologies. With gas stoves, the fire heats the air underneath a pan, and that hot air then heats the whole pan (as well as the handle, the stovetop, and the air in the room), which then heats the food. That's three separate transfers of heat, with each step adding waste and diminishing the amount of heat left to pass up the line, which slows down the cooking and requires extra time and energy to get the job done. That's what makes gas stoves the least efficient cooking technology. Conventional electric stoves are more efficient because they cut the process down to two transfers: the hot coils touch the pot directly, transferring heat to the pot, which then heats the food. Induction stoves do better still because they need only one heat transfer, pot to food, achieving 90 percent energy efficiency—about 10 percent more efficient than conventional electric burners and three times better than gas cooking. An induction stove doesn't make any heat to transfer, and the "burners" aren't burners at all—that's just a term of convenience. The ceramic glass cooktop, like the book in that 1933 World's Fair demonstration of cold cooking, is unaffected by magnetism—so even the part directly under the pan doesn't get very hot, except from heat reflected back from the bottom of the cookware. A metal pot handle won't get uncomfortably warm on an induction stove, either, because the magnetic field reaches only the bottom of the pot. On a gas stove, you better use a potholder to avoid painful burns.

Different but Better

Still, Galarza wondered: Was all this too good to be true? There had to be some trade-offs. One obvious shortcoming was that only pots made of ferrous material—metal that magnets stick to—work with induction. Cast iron, steel, steel alloys, enameled cast iron, and any pot with a steel-clad bottom will work. Copper and aluminum without that steel bottom won't. He knew that wouldn't be an issue for the vast majority of chefs, who generally don't use anything but sturdy iron and steel. But it could give pause to home cooks who might have inexpensive aluminum or pricey copper-clad cookware. Then there were a few cheffy things that took some getting used to with induction. The toughest was learning cooking without the visual cues from a flame, which let you set temperature by sight. He also initially missed twisting and tweaking a familiar stove knob, which made it easy to set to the right position without even looking through feel and muscle memory. Induction meant setting a specific temperature with a button and a digital readout. Some manufacturers have begun adding old-school knobs and blue lights on the cooking surface to simulate flames, but it's still very different and takes some getting used to. And you can't just toss a pepper directly onto a burner to char (though pan charring was so fast with induction, Galarza thought, it didn't really matter much).

Complaining about such minor details, in Galarza's view, amounted to standing on the bridge of the starship *Enterprise* and complaining that you missed the romance of steamships like the *Titanic* with their coal-fueled boiler rooms and soaring smokestacks. Nostalgia had its appeal, but Galarza decided induction was better, safer, and cooler (in both the literal and slang sense of the word) than gas cooking. Safety was a significant differentiator for Galarza. Burns are a frequent injury in commercial kitchens, costly in terms of time, money, and staff health. And in a world where forgotten or poorly tended stove burners are the number one cause of home fires, a stove that doesn't actually have hot surfaces is a game changer. An induction burner with no pot on it is just a glorified refrigerator magnet that will automatically shut down, a built-in fire-safety feature.

If there were any lingering doubts, they vanished when he ran the time-saving numbers, which were astonishing for any busy restaurant chef. Water boiled twice as fast. An egg fried in less than half the time. Then Galaraza read a cooktop comparison study that found, based on the time it took for an eight-quart pot to come to a boil, that a high-end gas stove can produce a fraction less than 39 pounds of cooked food an hour, while an induction stove can cook about 71 pounds in that same hour. Time is money in restaurants. This boost to productivity could shift the ground from financial struggle to profitability in an industry where margins run razor thin. There's a reason why time (and time running out) is an ongoing theme in the second season of the hit TV restaurant drama *The Bear*, where the kitchen wall clock has a sign below it in all caps: EVERY SECOND COUNTS.

There's more than just time saved with an induction stove. A large commercial kitchen typically needs a massive and powerful ventilation and exhaust system to cleanse the indoor space of smoke, particulates, and toxins caused by flame cooking. Such a ventilation system for a gas-powered kitchen the size of his at Eden Hall would have cost Galarza about $100,000. For his new electric kitchen, however, the ventilation needs were smaller. Unlike conventional gas cooking, in which pans are kept stacked on burners preheating to save time, and broilers and ovens are always on, preheating with induction is unnecessary. Galarza put this claim to the test by sticking a favorite pan in the freezer. When he pulled it out and put it on the induction stove, it was cooking at searing temperatures in seconds. Yet instead of installing a cheaper ventilation system for his cool and toxin-free kitchen, he bought an expensive smart system with sensors that constantly adjusted ventilation, from high to low to off, cooking station by cooking station. This cost $300,000, three times as much as a comparable conventional system.

Later, when he told his friends and colleagues about this, the other chefs were agog. Why would he do that? Was he crazy? He just smiled and told them what had happened with Eden Hall's electricity bills. In gas-powered kitchens, those ventilation systems run all the time, and the electricity costs a fortune. The savings from his electric bills for his new smart system made up the extra $200,000 upfront cost in just nine months. After that, he saved money every month.

"Nine months!" Galarza says. "That system has paid for itself many times over."

When restaurant owners hear that, more than a few want to know one thing: When can he help redesign their kitchens, too?

The final selling point for Galarza was the environmental impact. The same factors that make induction cooking cheaper, faster, and more energy efficient than cooking using fossil fuels also make it less wasteful: it eliminates the heat, the toxic pollution, and the greenhouse gases that are all waste products of gas stoves. Gas stove emissions include two greenhouse gases: carbon dioxide from burning gas, and raw methane leaked whether the stove is on or off. Methane is eighty-seven times more potent than carbon dioxide as a greenhouse gas, which means the leaked methane alone from the nation's gas stoves is the environmental equivalent of a half million gasoline-burning cars—annually. Stoves are one reason that the nation's buildings are a major driver of climate disruption, accounting for nearly 13 percent of all heat-trapping pollutants.

Bringing Induction Cooking to the Masses

Galarza realized that he and other chefs could use induction cooking to build safe, healthful, and profitable businesses while also being good stewards of the planet. Better kitchens, better food, better service to customers, better conditions for the cooks, and a way to help save the world, too—the electric kitchen won on every level. And he realized that his position at Eden Hall wasn't just a cool new job; it was life-transforming. Galarza had found his calling. He would not only give up gas burners for good but would launch a new consultancy, Forward Dining Solutions, dedicated to electrifying American cookery, one pilot light at a time. Cooking with magnetism sells itself, he enthused to anyone who would listen: "Just try it. It will blow your mind."

The timing of Galarza's new business launch couldn't have been better. It coincided with a decarbonization movement emerging in many American cities and states, aimed at making the built environment—our homes, businesses, schools, and government buildings—less wasteful and polluting by converting all systems to electricity. It didn't hurt that his first big client was Microsoft. The tech giant's sustainability staff wanted an all-electric dining commons like Eden Hall's for its sprawling campus in Seattle.

Galarza didn't know that this movement would eventually ignite (yet another) front in the culture wars. At first, the pushback came mostly from the restaurant world. He wasn't surprised. He was trying to overturn traditions and habits practiced at every cooking school and by nearly every top chef in the country, and he threatened something deeper, too: the enduring appeal of cooking over an open flame evokes an emotional response that's hard to counter. The reaction from other chefs at first ranged from incredulous to eye-rolling to outrage at his treason. But he would show them his electric bills. He would demonstrate his induction cooking. He'd show them the "frying pan from the freezer" trick, which never failed to impress. For professionals who knew exactly what to expect from their gas stoves, and who had gotten used to the heat and discomfort and sweat-drenched tunics, Galarza's cool, clean kitchen was eye-opening. A friend and fellow chef in Pittsburgh, who was building (and has since opened) a new restaurant named EYV (Eat Your Vegetables), wasn't hard to convince. He had tried and liked induction cooking during an apprenticeship in New York, but he was completely sold when he realized he could save tens of thousands of dollars by building an all-electric kitchen from scratch.

"For me, it was an economic no-brainer," recalls Chef Michael Godlewski. "They asked me where I wanted the gas line, and I said, 'Nowhere.'"

Little by little, other chefs began to come around to Galarza's point of view. But things really started to click when he was joined in this sometimes lonely battle by a co-conspirator across the country, who worked on the home side of the cooking and appliance industry in the San Francisco Bay Area, and who had been thinking along exactly the same lines as the restaurant chef.

Home Ranges

Back in the 1990s, Chef Rachelle Boucher was working in fine dining when a tempting job offer came along: Would she be interested in a private-chef gig in Northern California's verdant Marin County?

Long favored by the rich and famous, Marin is fertile ground for private chefs. Legendary Hollywood couple Marilyn Monroe and Joe

DiMaggio called it home, as did music legends ranging from Tony Bennett to Jerry Garcia, Janis Joplin, and Bonnie Raitt, along with a coterie of billionaires who have made it one of the ten richest counties in America. Even celebrity chef Julia Child lived there as a teenager while attending an elite prep school. Boucher was not so easily dazzled—she had cooked for famous restaurant patrons before—but there was one Marin County notable Boucher could never refuse, given that she is a diehard Star Wars fan who owns, among other paraphernalia, several light sabers. So when a recruiter dangled the prospect of becoming the private chef for director-producer-screenwriter George Lucas and his family, just as the filmmaker was preparing to make his fourth Star Wars movie, *Episode I: The Phantom Menace*, it was a no-brainer. She said yes, and for the next three years cooked for the Lucas family at their home, at elaborate events hosted by the filmmaker, and on the road.

The kitchen that awaited Boucher in Marin County was, as she suspected it would be, beyond deluxe in design and equipment. The star of the show occupied center stage: the high-end gas range that soon became her favorite toy. This was the 1990s, and cooking with gas was almost every chef's preference, as it had been for half a century at that point. To Boucher, it wasn't just that gas was traditional and familiar. It also was seen as the finest tool to make the finest food. As for its effects on the health of people and planet, hadn't the industry branded it *natural* gas, calling it the "clean" fossil fuel, the one that didn't spew smoke like wood and coal, nor noxious exhaust like cars? That signature clean blue flame seemed to signify purity. And nowhere is this perception more widespread than in California, where 70 percent of homes have a gas stove—the highest market dominance of any state in the nation. Most Americans cook with electricity. Only five states and Washington, DC, have more gas stoves than electrics. But in California, gas is king.

Boucher never questioned this until her next career change, when she took a job as a corporate chef, running workshops and managing events in the showrooms of luxury kitchen appliance makers. There she got to try out and demo the latest and greatest stoves, ovens, cookers, and cookware for potential customers, executives, buyers, and industry leaders. And it was here in 2013 when she cooked with induction for the first time.

Only a couple of induction ranges were on the showroom floor then, tucked out of the way, collecting more dust than sales. They had zero marketing, so customers rarely asked about them or even knew what they were. But Boucher was curious and decided to play with them. As with Galarza, she needed a few tries to get the hang of induction cooking, but in the end, she was smitten. Induction was everything Boucher wanted in a stove, the best qualities of electric and gas cooking combined, with none of either technology's drawbacks: Faster than gas. More efficient than conventional electric. Safer than both; if she forgets to take off one of the scarves she likes to wear, an induction stove can't accidentally set it on fire when she leans over the cooktop.

When she realized she preferred induction to gas cooking, Boucher wasn't thinking about the environment, or health, or climate disruption, or energy efficiency, or the fact that the powerful and expensive ventilation systems required in gas-powered kitchens would be obsolete with this technology. In time she would consider all this, and that would cement her allegiance to magnetic cooking for good. But initially, it was strictly because induction worked better. Her food turned out better. She likened it to moving from an old Volkswagen Beetle—or perhaps a horse and buggy—to a state-of-the-art EV. Even though those few induction stoves in the back were the oddball outcasts of the showroom, Boucher found she gravitated to them whenever she was there, using them for cooking demos and classes whenever possible. They became her go-to.

"The funny thing is, we had four chefs working there at the time, and without knowing what the others were doing, we each had made those induction ranges our favorites. We had all decided separately they were the best."

The problem was, the manufacturers and showrooms still wouldn't push induction stoves, despite the feedback they were getting from their in-house pros. Induction had caught on in Europe and Asia by then, where gas cooking was less common. But in the US market induction remained dead from neglect, except for a brief push as a hedge against gas stove pollution regulations that were under consideration in the late 1970s and early '80s but never came to pass. These few examples in the back of the showroom were the remnants of that effort. Because there were so few sales, prices remained high—not because the materials that went into an induction stove were more rare or expensive than those in conventional electrics, but because there was no economy of scale for the technology. And the expensive high end of cooking, after decades of vigorous marketing, was dominated by gas stoves. Electrics were supposed to be the cheap option. So induction was going nowhere fast in America.

When Boucher would ask the appliance company executives she met on the job why they didn't tout induction as the next great thing, they made excuses about price barriers or lack of demand or the challenge of introducing such an unfamiliar technology to replace perfectly good and familiar stoves. Yet at the same time, many execs admitted to her that they had ripped out their own gas stoves at home and put in induction models. *It's the best*, they'd confess to her. They were almost sheepish about it. Boucher started calling this the "clean little secret" of the appliance industry: The bosses knew induction was superior. They just didn't tell their customers.

Boucher found this attitude hard to fathom. Had Steve Jobs bowed to the same critiques with the first iPhone—that the tech was too unfamiliar, the learning curve too steep, the lack of a physical keyboard too scary— we'd have missed out on the most successful and disruptive technological shift so far this century. Instead, Apple invested in a huge marketing campaign and used the company's network of retail stores and Genius Bars to show new customers how to use the unfamiliar tech. Boucher wanted the appliance industry to do the same thing for induction, to get out the word on its virtues, and the technology would sell itself. Boucher argued— unsuccessfully at the time—that home cooks just need to be "Genius Barred" into falling in love with the cooking equivalent of the iPhone.

The funny thing is, the appliance industry had faced this problem before. The first gas and conventional electric stoves had to overcome strikingly similar barriers. Although they were introduced in the 1800s, they didn't catch on for decades, despite their many advantages. Consumers worried about and even feared replacing their familiar wood- and coal-fired stoves, which had been around for generations, and manufacturers didn't want to cannibalize existing sales. It wasn't until the 1920s and '30s that this dynamic changed because of outside forces: electric and gas utilities had gone to war over which would power the homes of the future. The two emerging energy superpowers launched national advertising campaigns touting the superiority of the rival stove technologies, hoping to boost sales of their competing fuel and volts.

Boucher longed for some similar sort of marketing lightning to strike on behalf of induction, and it finally arrived in the form of the building decarbonization movement, though this was a campaign to use less fuel, not more. And that put induction in a starring role at last.

In 2019, Berkeley, California, became the first city in the country to launch an all-electric building policy, banning gas hookups in new construction. This San Francisco Bay Area city was already on the front lines of the war on waste. Building on one of the nation's oldest and most successful recycling programs, in recent years the city has outlawed disposable foodware for indoor dining, required compostable containers and no plastics for takeout meals, and slapped a twenty-five-cent fee on disposable single-serve cups to encourage reusables. The gas ban in new buildings didn't affect existing appliances, but it meant there would be no new gas furnaces, water heaters, or stoves added in the future. More than seventy-five other California communities followed suit, Los Angeles and San Francisco among them, plus dozens more around the country, including Denver, Seattle, and Boston, as well as Washington and New York States. Washington, where 84 percent of homes already had electric stoves, chose to limit its gas ban to furnaces and water heaters.

The rise of this building electrification movement was the sign Boucher had been waiting for. She left her corporate chef job to begin the latest phase of her career, as consultant, advocate, and evangelizer for kitchen electrification. Now she runs culinary events for the California-based Building Decarbonization Coalition, a nonprofit working with industry and communities on home and business electrification. She and Galarza met through this group and decided to work together. They use webinars, live cooking demos, conferences, legislative testimony, and social media to talk up the advantages of induction cooking, and the pair collaborated on a kitchen electrification practice guide published by the coalition.

Her business's name is Kitchens to Life, a reflection of her conviction that getting fossil fuels out of the kitchen (and the whole house) is good for everyone and adds to the joy of cooking, rather than dishing out a foodie buzzkill. She bases this on decades of peer-reviewed scientific research, stove performance tests at the renowned Food Service Technology Center in San Francisco, and her professional and personal experience, including her own husband's struggles with asthma attacks in the presence of the gas stoves she once happily used. She believes that the health issues that result from gas cooking alone are reason enough for everyone to ditch cooking with gas, no matter how they feel about climate change or electrification.

"It's night and day, and it's real," Boucher says. "My husband hasn't had an asthma attack since we switched."

The Reality of Gas and Health

Learning about the health effects of gas stoves hit Boucher hard. She had always believed it was the clean fuel to use, a healthy way to cook. That's what she was told, she says. It's what everyone was told. The gas industry is still promoting that fiction. But the science is clear, Boucher says: Those gas stoves she long loved emit airborne chemicals in the house. In sufficient quantities, they can make you dizzy. They can make your heart race and your breath feel short. They can make you sick, attacking your lungs, your heart, your brain, and your blood. And under the right conditions, they can kill. They *have* killed.

Besides the heat-trapping pollutants methane and carbon dioxide, stoves emit poisonous carbon monoxide, particulate pollution, asthma-triggering nitrogen oxides, and the carcinogens formaldehyde and benzene. This is virtually the same toxic brew that comes out of car exhaust pipes in greater quantities—the stuff that kills you if you run your car with the garage door closed. And the connection to childhood asthma and respiratory ailments has been documented in scientific studies for over fifty years.

THE FACTS ABOUT HARMFUL
GAS STOVE EMISSIONS

Carbon monoxide is an odorless, colorless, and poisonous waste gas. Breathing it in sufficient concentrations causes death by blocking the blood from absorbing oxygen. It causes you to suffocate in a room filled with air. The more common risk for most people, however, is long-term low-level exposure, which can take a slow and imperceptible toll, causing chronic fatigue, flulike symptoms, memory loss, musculoskeletal pain, mood swings, depression, and, finally, permanent brain damage. Accidental carbon monoxide poisoning causes more than four hundred deaths and around one hundred thousand trips to the emergency room every year in the United States. Malfunctioning and poorly maintained gas furnaces and stoves are frequent culprits. So is the potentially lethal practice of using an open oven for heat in the winter.

Particulates are charred particles, often tiny enough to be measured in nanometers, that are generated by burning fossil fuels. Because they are so small, they can penetrate deeply into lung tissues and cause chronic inflammation.

Nitrogen oxides (collectively called NO_x) irritate the lungs and raise the risk of lung disease and asthma. Children living in a home with a gas stove have a 42 percent greater risk of developing asthma symptoms, according to a groundbreaking study in 2013. More recent research on how often that risk becomes reality has found that nearly 13 percent of all childhood asthma cases in the United States were caused by gas stoves, according to a joint study by RMI (former Rocky Mountain Institute), the University of Sydney, and the Albert Einstein College of Medicine. States with the highest percentage of gas stoves in homes exceeded 20 percent.

Childhood Asthma Cases Associated with Gas Stoves

State	% Asthmatic Children Linked to Gas Stoves	% of Homes with Gas Stoves
ILLINOIS	21.1%	67%
CALIFORNIA	20.1%	70%
NEW YORK	18.8%	62%
MASSACHUSETTS	15.4%	44%
PENNSYLVANIA	13.5%	37%
TEXAS	11.7%	37%
COLORADO	10.8%	31%
OHIO	9.5%	34%
FLORIDA	3%	8%
USA TOTAL	12.7%	38%

Data sources: RMI, US Energy Information Administration

Formaldehyde and benzene are classified as carcinogens, with benzene particularly virulent. Benzene is a known cause of leukemia and other blood and bone marrow diseases in humans, according to the US Department of Health and Human Services and the Centers for Disease Control and Prevention. Homes with gas stoves can reach levels of one or the other that are high enough to be considered unhealthy, though there is no truly safe level of exposure. Tests of the air inside homes with gas stoves in use show benzene levels greater than from breathing secondhand cigarette smoke, and sometimes matching the fumes that create hazards around gasoline pumps, where cancer-risk warning signs are typically posted, though few customers pay attention. The benzene emitted by gas stoves while cooking food was ten to twenty-five times more than could be detected around old-school electric stoves, and remains in the air for up to six hours after the stove is shut off. Induction stoves had no measurable amounts of benzene.

"I'm hard-pressed to think of a more powerful chemical cause of leukemia than benzene," oncologist Jan Kirsch told *The Guardian* newspaper after this latest study was published in 2023. "People have died, undoubtedly, from exposure to benzene in their homes and unless this problem is ameliorated people will continue to do so."

Rob Jackson, a Stanford University professor whose work on gas stove emissions includes revealing the colossal methane leak problem, says the science should be a gas stove deal-breaker for anyone who cares about their own or their kids' well-being. "I don't want to breathe any extra nitrogen oxides, carbon monoxide or formaldehyde. Why not reduce the risk entirely? Switching to electric stoves will cut greenhouse gas emissions and indoor air pollution."

These harmful chemicals are unavoidable when burning natural gas for any purpose, which is why every other gas-powered appliance in the home has *always* been vented to the outside through a flue, chimney, or some sort of exhaust port: gas furnaces, hot-water heaters, even gas-powered clothes dryers all direct 100 percent of their pollutants outside the house. Stoves are the only exception, even though their emissions with all the burners lit and the oven on can exceed those of a hot-water heater and match those from a furnace powerful enough to heat a 1,200-square-foot home.

This wasn't always the case. Early gas stoves had flues, just like their coal and wood competitors. As later models became streamlined and built-in, rather than the old stand-alone behemoths, the flues went away, as the myth took hold that gas stoves were somehow "clean," when no other gas devices were. In the 1970s, range hoods with exhaust fans positioned over the stove became more common and are now widespread, though far from universal. Their effectiveness is questionable. Many of them are not vented to the outside by a chimney or duct, but simply blow the pollutants back into the house—reducing them to useless noise-makers. The main problem with range hoods, though, is their fundamental design: they have to be turned on. And surveys show that most people, most of the time, don't bother.

Does this mean you or your kids will definitely fall ill from cooking with gas? No. This is like asking, *If I drink and drive, will I definitely crash?*

Or, *If I'm exposed to a virus, will I definitely get infected?* Living with a gas stove is a risk, not a guarantee, because courting disaster is not the same as creating it. But with time and repetition, any risk always grows, moving from unlikely to even money to the brink of certainty. No risk in your home is more repeated, constant, or long-lasting than the stove you use day in and day out. And the risk is higher still for the most vulnerable people in the house: the elderly, lung or heart patients, children, and babies.

There are two ways we can regulate these sorts of safety issues in which ordinary people are unwittingly exposed to harm by a product or process. One is the method of proving causation, usually through a lawsuit and trial. This comes into play when someone believes a product has made them sick, or killed a loved one, and to prevail, the person injured has to prove that their ailment was actually caused by the product. This is hard to do. Asbestos-exposure victims who developed a rare cancer were able to meet this burden of proof. People downwind of the Three Mile Island partial nuclear meltdown in 1979 who developed cancer were not able to prove definitively that the radiation from the disaster was responsible. It's unclear where the gas stove would fall on this spectrum.

The other method is the precautionary principle, which takes the opposite approach when a product exposes people to conditions—such as chemical emissions—that create a possible risk of harm. Under the precautionary principle, the manufacturer is required to prove that no harm will come from such exposure. It shifts the burden of proof in favor of *precaution*. In a world where every other gas-burning device in homes, businesses, and industries is regulated to avoid indoor emissions because of their harmful properties, it seems unlikely that the gas stove could be proved to be harmless in its current form. Induction stoves, however, would pass the precautionary test with flying colors. The precautionary principle tells us that, all other things being equal, choosing the least risky option always makes sense.

"Some risks are unavoidable. But gas cooking isn't one of them," says Boucher. "When we've got so many amazing alternatives now, there's really no justification for the risk. It's just not worth it."

WHAT YOU CAN DO TO REDUCE THE RISK
OF COOKING WITH GAS

Chef Boucher hates scaring people, so she and others involved with limiting indoor air pollution suggest a list of steps that can reduce harm, short of running out and buying an expensive new stove, which is not an option for everyone.

- Cook with the windows open. Let the fresh air push out the waste gas pollutants.

- If you have a range hood that vents to the outdoors, always use it (preferably with the window open, too). Try to do most of your cooking on the back burners, where the range hood is most effective.

- Close bedroom doors and other spaces to keep most of the toxins out of those rooms while cooking.

- A good window fan set on exhaust can also help.

All these strategies can make a big difference. But even smaller doses of carcinogens and poison gas are still risky, so the safest course is to avoid cooking with gas as much as possible.

- Get a single countertop induction burner. This is an affordable alternative that gives you a chance to try out induction for about $100 (less if you shop sales). Good product recommendations can be found on websites like Wirecutter, Tom's Guide, and Consumer Reports. This can become your main cooking tool. Many restaurants use them for their speed, price point, and flexibility to create a workstation anywhere, anytime. If you already have a toaster oven or an electric slow cooker (or both), adding an induction burner to the mix can fill almost all your cooking needs without ever firing up the built-in gas stove. You can reserve those blue flames strictly for big feasts and special occasions—a backup instead of a mainstay.

- Use federal incentives payments from the Inflation Reduction Act of 2022 to cover part of the cost of a new induction range. If you are a homeowner (or a landlord), induction ranges and other high-efficiency appliances are eligible for significant rebates.

5

Taking the Heat

As the push for electrification and the gradual phasing out of gas in buildings gained momentum in the United States, and several new studies came out suggesting the health and environmental risks posed by gas stoves were far more alarming than previously believed, induction shifted from obscurity to trend. Cooking with magnets started to explode—in the news, in social media, in the demand for Galarza and Boucher as speakers and consultants who were frequently quoted in the press, and in the apartments of millennials and Zoomers, who discovered the rapidly growing line of portable, affordable high-quality induction cooktops flooding the market. Full-size ranges of any kind are expensive, with the cheapest models in the $500 range and the greatest selection bumping $1,000. The switch from gas to electric ranges can add yet more cost, as it can require some wiring upgrades in older homes. But the countertop portables—which can be quite powerful single- or double-burner induction cookers—are more like the cost of a dinner and a movie for two. They were priced to move. And they were moving, with some models racking up thousands of favorable online customer reviews.

It made sense: They are perfect for students, singles, and apartment dwellers whose space is at a premium. They are also a way of keeping a gas stove but making it only a special-occasion polluter, rather than a daily workhorse.

TikTok star and chef Jon Kung of Detroit began posting videos and cooking tutorials for his tabletop induction burner to his 1.5 million followers. In one video, while he cooks a sizzling fried rice and vegetable lunch, Kung explains his own evolution from gas cooking to induction, the steam rising from the pan as he chops and adds ingredients: "A few years ago, if you asked which was my favorite heat source to work with, I would have said gas. Probably ninety percent of the cooks out there would say the same. . . . That changed when I learned about induction burners. I started with just one plug-in cooktop. I had it for years, and I used it frequently when I did my pop-ups. . . . I started using it more and more, expanding my collection to the point where they are now pretty much all I use. 'Oh, cool, I'm cooking with magnets,' I thought."

For Kung, the speed of induction cooking took some getting used to—he had to react more quickly to avoid overcooking or burning. And induction can warp pans if you try to preheat them empty, though he quickly learned preheating wasn't necessary. The unfamiliar cooking speed, unsurprisingly, became an advantage rather than an obstacle once he adjusted to it, because his meals and pop-up menu courses could be made so much faster. Aside from its speed, his portable induction burner could be used at indoor venues where his unvented portable gas burners could not, and his own small kitchen stayed cool because of induction's efficiency. He was sold on the technology before he even knew about gas stoves' indoor air pollution problem.

"It only reinforced my thinking to make the full switch," he tells his audience. "Honestly, it wasn't much of a sacrifice. It's only made my life easier in the long run."

The Backlash

To the gas industry, the proliferation of such endorsements from influencers posting for a young foodie audience was a five-alarm fire. A tidal wave of pushback and, at times, misinformation rolled into the mainstream and social media in response. The gas industry hired its own influencers to demo the joy of cooking with real fire, and trolls started commenting on induction advocates' posts and articles on the subject. The "woke police" were using fake science to tell you how to cook, some griped. Others falsely claimed that induction stoves' magnetic fields caused headaches and made pacemakers go wild. And some posts made inflated claims about high electric bills, while others argued that gas is better for the environment than renewables, that it's a "bridge" needed to get us safely to an all-renewable future, or that it actually *is* renewable. None of those claims are true.

Anger initially focused on the first mover on this issue, the city of Berkeley. Critics didn't want to talk about electrification or decarbonization. They called it a "gas ban," with some chefs and restaurateurs unfamiliar with induction deploring electric cooking as inferior. Some in the Bay Area's vibrant Asian-cuisine scene worried about the effect on their traditional culinary techniques of cooking with fire, notwithstanding induction's popularity in the actual home of Asian cooking, Asia.

The powerful California Restaurant Association sued the city. The organization claimed that Berkeley overstepped its authority, misrepresented the science, and was killing businesses and jobs by assaulting choice, traditional values, and cultural practices. The gas industry wasn't a party to the suit, but the state's largest gas company had donated money to the restaurant association, and the industry talking points were liberally quoted in the lawsuit. The conservative magazine *National Review* made it a partisan issue: "The Democrats' War on Gas Stoves Is a Slap at Cooking Cultures."

Meanwhile, one community after another adopted variations of Berkeley's all-electric building policies, some out of concern over indoor air pollution, others in pursuit of sustainability and climate goals. The fossil fuel industry tried to attack the scientific research on gas stove hazards, but the most visible expert it hired was not taken seriously. She had offered similar defenses in the past for Big Tobacco and the plastics industry, argued against regulation of toxic air pollutants such as mercury, and cast doubt on widely accepted research linking air pollution to reduced life spans. Her credibility as a hired gun didn't matter to the gas industry, though. She served her purpose, creating the appearance, if not the reality, of controversy and uncertainty.

Then a commissioner with the US Consumer Product Safety Commission offhandedly remarked during a long interview with a reporter that one option to make gas stoves less hazardous would be to impose stricter limits on emissions. And if that didn't work, he pointed out, the commission had the power to ban the sale of any unsafe product. It had famously done so decades earlier with a terrible and deadly toy, lawn darts, after the sharp metal projectiles injured more than six thousand people and killed two small children. The only question then wasn't *But what about our freedom of choice?* It was *What took you so long?*

Stoves, though, are not lawn darts—with their rather obvious flaw of making a child's throwing toy out of a large and aerodynamic ice pick. The gas stove's flaws and dangers seem far more remote, if not invisible, while its traditional virtues seem abundant, not to mention shielded by the stubborn power of "But that's the way we've always done it." The commissioner's remark was clearly just innocent speculation, not an actual plan, policy, or proposal. But it was twisted into the spark the fossil fuel industry and its allies had been waiting for. They pounced, and willfully misrepresented the comment as a promise and a threat by a federal Big Brother agency taking aim at a cherished product. This, the industry said, was an attack on Americans' freedom of choice.

"If the maniacs in the White House come for my stove, they can pry it from my cold dead hands," one congressman declared (from a state where most people cook with electricity). Another, also from a majority electric-cooking state, posted on social media with his own alliterative holy trinity: "God. Guns. Gas Stoves." Then the governor of Florida, a presidential hopeful from a state where hardly anyone cooks with gas, vowed to give tax breaks to encourage gas stove purchases, which most of his Sunshine State constituents couldn't possibly use. Meanwhile, twenty states passed laws banning their own cities from banning gas, with one, Oklahoma, also making it prohibitively costly for consumers to disconnect gas service on their own, which is an odd way to defend freedom of choice—by taking away choice. The US House of Representatives got in on the act by passing several hastily written bills to defend the grand tradition of gas cooking—legislation that had exactly no chance of becoming law, but which added fuel to the fire. Finally, *The Wall Street Journal* cheered this on with a column decrying "extremely powerful climate groups" that will not rest until they have "dictated what you drive, where you live, and how you cook."

Suddenly, the most antiquated technology in our homes had been elevated to a cause worth fighting to the death to defend—from a completely made-up threat. But that's how you turn an innocuous conversation about consumer product safety into the opening salvo of a new front in the culture wars.

It's hard to say which part of this series of events is most absurd. One contender for the silliness crown is the claim that there exists such a thing as "extremely powerful climate groups" capable of pushing around the mightiest, most profitable, and most lobbyist-laden industry on the planet. If there were such a unicorn out there, we all would have been driving electric cars and powering much of our world with wind and solar by the year 2000. That was the plan announced in 1979 by an environmentalist president, Jimmy Carter, who set that goal during the height of the 1970s oil crisis. That plan for a new century of energy independence, forged as a matter of national and economic security, was happily canceled by his successor, Ronald Reagan, who allowed Carter's renewable energy tax credits to expire on December 31, 1985. A few months later, the solar energy system Carter proudly put atop the White House was taken down and stuffed unceremoniously in a warehouse (though its thirty-two solar panels were taken out of mothballs and reused in later years on a small college campus in Maine, with one panel put on the roof of the Carter Library in Atlanta). Those "extremely powerful climate groups" are so influential, it took twenty-eight years to get solar back on the president's home, when Barack Obama had new solar panels installed in 2013 to fulfill a pledge he made shortly after taking office.

Then there's the simple fact that no one who meets Chef Rachelle

Boucher or Chef Chris Galarza could mistake their inclusive, optimistic evangelism for induction stoves as dictatorial or oppressive, or about anything other than giving people the information and tools to choose the best way to cook for them. It's just not credible—but only if you know them.

The biggest lie, though, was the pretense that there was a plan to seize people's existing gas stoves. No one has ever proposed such a thing. Nor does any community or agency have the power, authority, or practical ability to do it. As if a country with critical shortages of cops, 911 dispatchers, and a host of other law-enforcement staffers has the workforce available for tens of thousands of crowbar-wielding stove-police squads to seize gas ranges from *47 million homes*. The very idea is ludicrous.

The only thing the electrification ordinances call for in Berkeley and the other places that followed its lead is an end to natural gas lines *in new construction*. The Consumer Product Safety Commission, if it acted at all, could indeed achieve the goal with even less effort simply by setting more stringent emissions standards for new stoves. For the past forty years, the gas industry has known that day could come. Its own internal research back then, along with a 1981 finding by the EPA linking respiratory disease to gas stoves, made it clear that emissions were a problem. The commission might be accused now of dereliction of duty for not acting on indoor pollution from gas stoves decades ago, but certainly no one can accuse it of a rush to judgment today. The uproar was based on nonsense.

Its effects, though, were anything but. This staged battle upset people. Just the prospect that some shred of it might be true was enough. Because Americans hate bans. They loathe being coerced. Even consumers who might have considered switching got mad, because, quite understandably, they want the change of something so integral to their daily lives and homes to be their own idea, and on their own timetable. For a general public not sure what to believe, this conflict hardened attitudes and divisions needlessly. As was intended.

The funny thing is, the gas industry doesn't really have any financial interest in the gas stove business. It was central to the industry back in the 1930s during the gas versus electricity wars, but in today's reality, where most natural gas sales involve the massively profitable business of generating electricity, stoves are little more than a rounding error. They represent less than half of 1 percent of gas sales, and the smallest portion of the total gas used in buildings. More than 93 percent of the gas consumed in homes is for furnaces and hot-water heaters. Even the combined category of clothes dryers, hot tubs, and pool heaters outstrips the gas consumption of stoves. The gains of ending the reign of the gas stove are even smaller for environmentalists: while natural gas burned for all purposes is responsible for a whopping 36 percent of all US greenhouse gas emissions, gas stoves contribute only two-tenths of 1 percent—another rounding error. That's why electrification advocates' main target in the home is electrifying heaters— space and water. That's where most of the waste, carbon emissions, and toxins caused by buildings comes from—except for indoor air pollution and its health risks. That's all gas stoves, though the industry disputes it.

And yet the fossil fuel industry decided to make this rounding error its Valley Forge anyway, recognizing what the electrification folks missed: stoves were the perfect wedge issue. Nobody is going to go to the mat over whether their heating system or their water heater is gas or electric. The only questions buyers of those big pieces of home infrastructure ask are *How much does it cost to buy, how well does it work, and what's the monthly operating expense?* You add the answers up, and pick the best product. Only stoves inspire the equivalent of brand loyalty to a fuel. They are unique. The marketing of gas stoves as the clean choice, the luxury choice, the kiss of fire, has been so effective that a 2019 survey showed only one in five gas stove owners would consider switching to another way of cooking. Many of us feel an emotional attachment to our gas stoves, tied to so many happy events from the past—Thanksgiving family dinners, baking cookies with Grandma, nostalgic aromas that evoke distant and pleasant childhood memories tied to those pretty blue flames . . . though none of the delicious smells and tastes change by replacing those polluting blue flames with induction. But the culture war over the gas ban struck a deep chord not easily swayed by rational arguments about the superiority of magnets versus fire.

The industry's hope was that the beloved appliances would be the gateway drug for keeping consumers hooked on gas for everything in the house—and beyond. So the Stove Wars were really a proxy war aimed at turning people against the electrification movement's overarching goal: decarbonizing not just buildings but the entire grid—the single largest market for gas, and the only significant one left for coal. This is the real existential threat that terrifies the natural gas industry. Losing the two thousand utility-scale US power plants that run on natural gas, the most of any country in the world, would put gas on life support.

Unfortunately, keeping that business model as is requires preserving the epic levels of waste, toxins, and heat-trapping pollutants and the wildly expensive inefficiency of gas-fired power plants, which, like giant stoves, waste a majority of the energy that they consume. Preserving the status quo is what the Stove Wars are really about: the hope that it's no big leap to go from declaring, *You'll have to pry my gas stove from my cold dead hands* to *Take your decarbonization and shove it.*

For a while, it seemed the fossil fuel industry had struck gold with this strategy. Then something surprising happened.

A City of Light, a War over Hot Air

A modern society powered solely by electricity is not a new idea. It was the dream in the late nineteenth century of the early titans of the electrical age—Thomas Edison, Nikola Tesla, and George Westinghouse. They showed off their vision, including an all-electric kitchen—stove, oven, hot-water heater, dishwasher, clothes iron, toaster, coffeepot—at the 1893 World Columbian Exposition in Chicago. Illuminated by more light bulbs than had ever been assembled in one place before—hundreds of thousands—the fair was dubbed the "City of Light" and the "White City." Dazzling spotlights lit the sky. The entire spectacle of daylight at night, with light bulbs outlining the walls and roofs of every building as if they were constructed of light, had enthralled a nation still lit elsewhere by flickering gas lamps, kerosene lanterns, and candles. More than 27 million people visited this world's fair during its six-month run, a staggering attendance at a time when the country's entire population was only 63 million. Most of the visitors had never seen an electric light before. It marked, in effect, the end of a different and very literal dark age, a mesmerizing vision of homes, government buildings, and businesses that needed no fossil fuels burning within. No one then worried about losing the wonderful tradition of burning stuff at home and work for lighting, heating, and cooking. Quite the contrary: They yearned to be freed from the flames, which represented the old ways. They wanted the City of Light, not the City on Fire.

This was not as close to becoming possible as many visitors believed. Although it represented a future so magical it is said to have inspired author L. Frank Baum's vision of the Emerald City in his 1900 book, *The Wonderful Wizard of Oz*, this City of Light was a demo and a prototype, neither practical nor built to last. It was as ephemeral as an *Oz* movie set, facades built of short-lived plaster and jury-rigged wiring. The early bulbs rushed into production just for the fair burned out so quickly that a full-time crew worked around the clock swapping in replacements. Neither America's tiny and localized electrical infrastructure of the era nor the technology behind it was anywhere close to being able to electrify everything everywhere out in the real world. Edison made his electricity the dirtiest and only practical way he had available—by burning coal for steam-driven turbines. But he viewed steam as a primitive and transitional method—he likened it to the "lumbering coach of Tudor days." He experimented with and even sold home generators that could be powered by wind, and saw the early work on primitive solar cells by a few of his contemporaries as the power source of the future. Shortly before his death in 1931, he reportedly remarked: "We are like tenant farmers chopping down the fence around our house for fuel when we should be using nature's inexhaustible sources of energy—sun, wind, and tide. . . . I'd put my money on the sun and solar energy. What a source of power! I hope we don't have to wait till oil and coal run out before we tackle that."

When the grid and the technology of power transmission finally matured in post–World War II America, fossil fuels were ascendant.

Burning fossil fuels in the home had been relentlessly marketed away from the prevailing view—that it was primitive, dirty, and old-fashioned—into a celebrity-fueled vision (featuring such midcentury Hollywood icons as Bing Crosby, Marlene Dietrich, Jack Benny, and even a cartoon Daffy Duck) that spun gas as the luxury, quality, clean-yet-affordable choice of the future. There were no latter-day Edisons or Teslas or Westinghouses commanding the spotlight to reignite the old all-electric vision. Ironically, it was Ronald Reagan, who as president would be a tireless booster for fossil fuel consumption, who was the most visible champion of the all-electric future as the paid spokesperson for General Electric, the company Edison had cofounded. In 1957, Reagan moved into the "General Electric Showcase House" in Los Angeles, also known as the "House of the Future." The house became a familar presence in prime-time TV advertisements as well as Reagan and his family's actual residence until he moved into the White House and stopped pitching electricity as the savior of us all.

Now, more than 130 years after the City of Light's demise and four decades after Reagan sold his GE wonder-house, that vision is being resurrected again out of mounting concerns over fossil fuel energy costs, toxic emissions indoors and out, and climate-disrupting pollutants that are cursing us with blistering summer heat waves.

But instead of the City of Light's aura of a new and better way to power the world, or the gee-whiz early space-age wonders of Reagan's showcase house, electrification is being wrongly cast as a killjoy out to end our best cooking traditions, our food culture, and our choice by taking away those iconic blue flames.

In turns out, though, that there's a dirty secret behind this defense of tradition: it's based on a lie. The story now told by the industry and largely accepted by lawmakers and journalists is that natural gas has been a major source of heat, light, energy, and flame for nearly two centuries. But the truth is, the stuff being piped into our homes for stoves and heaters and clothes dryers was a mere niche product that wasn't in wide use in America when gas stoves were invented and became popular. Indeed, what we are currently burning in our homes did not achieve market dominance until the 1960s and '70s. Before then, the stuff that most people burned wasn't even called "natural gas," a warm and fuzzy label for something that is "natural" only in the sense that volcanic lava, toxic arsenic, and radioactive uranium are "natural." You wouldn't want any of them in your kitchen, either.

The fuel that the phrase "now we're cooking with gas" referenced when it was coined in the 1930s, and that established our current tradition of cooking with open flames in the American kitchen, was not "natural gas" at all. It was a very different *manufactured* product made from coal that consisted of a mixture of hydrogen and carbon monoxide, called town gas or coal gas at first, and eventually just gas. It was developed exclusively for streetlights and then interior house lighting. Later, when electricity began to corner those markets, gas stoves were trumpeted so that the gas companies would still have a consumer business.

This gas smelled, burned, and polluted differently than today's natural gas, lacking the potent heat-trapping pollutant methane that today's natural gas is made of. The old gas was the stuff that people fell in love with and embraced as a good way to cook, and that the fossil fuel industry scrapped for a more profitable alternative much more recently, something pumped from deep underground rather than baked out of coal.

So it's not environmentalists out to kill a traditional way of cooking. The fossil fuel industry itself already did that, and it's now fudging that history so that it can claim induction and electrification advocates are the real threat to tradition.

As for those blue flames everyone professes to love? The funny thing is, they're just a light show. That blue light doesn't cook anything. The main output of burning any fuel is invisible infrared radiation, otherwise known as thermal energy, otherwise known as *heat*. And whether the source is burning natural gas, heating coils on an electric stove, or the energy transfer of induction, it's all just infrared. It cooks the same. The laws of physics don't care if fire is pretty, and neither does the steak or cauliflower or seafood paella you're cooking. Heat is heat, and flames are just a pretty waste of energy and clean air. The Stove Wars are really just a lot of hot air.

Where the Gas Industry Is Losing Ground

Buildings are America's fourth-largest contributor to climate disruption. But they are also the easiest slice of the climate pie to fix, because the latest generation of electric heating systems, water heaters, and stoves are all remarkably clean, efficient, less costly to operate, and low waste. And even with today's electric grids, 60 percent of which are powered with fossil fuel–driven generators (a percentage that drops every year as more renewables come online), electrification in buildings significantly lowers their footprint. This is what propels the building decarbonization movement nationwide.

Stoves may be the center of attention and controversy, but the real driver for electrifying everything is another new take on old science, though it's not magnetism this time. It's the heat pump, which has its own interesting twist: it's yet another way to heat things without making heat. And the bonus is that heat pumps can shift into reverse and make your home cool, too. It's two expensive pieces of infrastructure in one, which means that for new construction and home remodels, heat pumps are a slam-dunk economic winner.

Like induction stoves, heat pumps burn no fuel to make heat—they just move heat around (which is why they are *pumps*). So how does that help us in winter? It may not feel this way, but even on the coldest winter day, there's a lot of residual heat—infrared radiation—to be harvested outside. If that wasn't true, the earth at night would be like the surface of the moon and reach absolute zero (–460 degrees Fahrenheit, or –273 degrees Celsius), making life on earth impossible. Heat pumps grab this residual heat with a clever use of something we all know from the behavior of water: evaporation, the process of a liquid becoming a gas. Evaporation absorbs heat from the surrounding area, which is why sweat cools our skin as it dries. The effect is more pronounced if you put a dab of rubbing alcohol on your skin, which feels cold to us. That's because alcohol has a boiling point much lower than water's and becomes a gas very quickly at room temperature, stealing heat from our skin just as quickly. We experience this as cooling, but in physics it's more accurately called "heat transfer." Heat pumps magnify this further with a substance called a "refrigerant" with a very low boiling point—stuff that normally wants to be a gas. First used in refrigerators, ammonia was the original refrigerant, but it's been eclipsed by various synthetics, propane, or simple carbon dioxide. It's used to cool the air inside the fridge by transferring heat to your kitchen; that's why there is a fan in back of the appliance and the air coming out feels hot. The magic of the heat pump is that it moves heat into the house the same way from *outside*. In winter, the gaseous refrigerant is pressurized, forcing it into a liquid state, then it's pumped outside and depressurized. It expands rapidly back into a gas and sucks heat out of the outdoor air like alcohol on your skin. Then this heated gas is pumped back into the building and pressurized back into a liquid state, forcing it to release all that newly acquired heat to warm the house. In summer, a bit more magic comes into play with the flip of a switch, which reverses the process and converts the heat pump heater into a super-efficient air conditioner.

Depending on what older heating and cooling systems it replaces, a new heat pump can cut utility costs and waste by up to half, and its carbon emissions within the home are zero. In Washington State, where much of the electricity is carbon-free hydropower, heat pumps are a complete win for pocketbooks, pollution, and climate, and a vast improvement even in areas with more fossil fuel–generated electric power.

Heat pump water heaters use the same principles, and together the two devices could cut the total energy use and climate impact of the average US home by about half.

And it's here where, to the gas industry's surprise, the Stove Wars' proxy battle against all electrification failed to have the carryover effect that fossil fuel advocates hoped for. Their proxy war collapsed.

Hostility toward phasing out other gas appliances never materialized.

Instead, sales of efficient, low-waste electric heat pump systems surpassed those of gas furnaces for the first time ever in 2022. Heat pump water heater sales are also rising. This is a sea change that knocks out the most wasteful, polluting, and climate disrupting tech in buildings, responsible for 93 percent of their heat-trapping pollutants. This is happening nationwide in states with or without bans. In fact, heat pumps are most popular in southern states, although New England isn't far behind, especially in chilly Maine.

California has the toughest approach on gas so far, ending the sale of all gas appliances, including furnaces and hot-water heaters, by 2030. But the lawsuit against the Berkeley ban succeeded in having it overturned—though that ruling is being appealed—and additional suits in defense of gas are brewing nationwide. Yet this battle over bans is looking increasingly like a rearguard action. The appeal of heat pumps is winning in the marketplace without them—and without any culture war angst. The biggest complaint new heat pump owners have is: *Why didn't somebody tell me about this ten years ago?*

As for gas stoves, consumer choice will decide the battle. The benefits of trading in nineteenth-century cooking tech are real. The issue of indoor air pollution is real. Most crucially, the health risks to children are real. That's the key to change on this piece of the electrification puzzle: kids, not bans.

Years ago Walmart decided to invest heavily in organic clothes and products as part of its initiative to become more sustainable. At first, when it pitched its new line of organic cotton clothing as good for the environment, sales were terrible. Organic products cost more, and Walmart found its main customer base might choose the more sustainable option if there was no added cost, but otherwise, no sale. But when the company started marketing organic babies' and children's clothing and products as healthier, chemical-free, safer alternatives, sales skyrocketed. Walmart became the leading seller of organics in the country. Walmart shoppers wouldn't pay a premium for the climate, but they would for the health of their kids. They would make the switch.

What would it take to drive mass adoption of induction stoves without a ban in places where gas stoves now reign supreme? All it might require is a simple education campaign to make family doctors and pediatricians aware of the latest research on stoves and the respiratory health of children and vulnerable adults. Good diagnosticians should already be assessing patients with respiratory problems or symptoms of asthma by asking what kind of stove is in the home.

If that becomes common practice, and doctors start suggesting that worried parents consider ditching their gas stoves to keep their kids from wheezing and missing school, the days of gas cooking will be numbered. If I had known it was a potential trigger for my son's asthma, that thing would have been history in a heartbeat.

6

Squeezing the Juice

Modern energy efficiency doesn't deplete a concentrated resource, like oil or copper. Made of ideas, it depletes nothing but stupidity—a very abundant resource.

—AMORY LOVINS

Never too chilly, never too hot, the Banana Farm (which is a house, not a farm) remains stubbornly, reliably comfortable no matter the season—defiant in the absence of any sort of furnace or air conditioner. Only the atrium next to the living room ever gets truly steamy, delivering the tropical lushness and higher temperatures needed to nurture the home's namesake banana crop, coffee plants, guava fruit, and assorted other growing things year-round.

A modern building with no sign of the burning, chugging, fuming, and roaring machinery of heat and cold is not normal, except perhaps on some idyllic island. It would be truly bananas to try such a feat in one of the coldest climates in America. But that's just where the Banana Farm stands.

This spacious single-story home is perched seven thousand feet above sea level in the Rocky Mountains, in Old Snowmass, Colorado, an area best known for its plentiful snowfall, skiing, and harsh winters. The weather dips below freezing for seven straight months of the year, with the average low in January around 9 degrees Fahrenheit and with half a month's worth of subzero days thrown in to keep things interesting. Yet the Banana Farm stays warm and sustainable passively—through canny design, high-efficiency appliances, smart siting and angles to leverage sunshine in winter and shade in summer, and world-class insulation of walls and windows (about five times what conventional engineering says is practical, and way more than the poorly insulated typical American house has). The building further reduces waste by using less than half the water and a tenth of the electricity consumed in the typical US house. Solar photovoltaic panels on the roof provide electricity for lights and appliances, and a passive solar thermal system heats water.

It has twice the window space of a standard house, and though heat buildup on sunny days would normally be a problem, its state-of-the-art windows are so well insulated that they are nearly as good as walls. Even so, the atrium's wall of south-facing insulated windows still admits enough heat, along with copious sunlight, for the tropical plants to thrive. Its purpose goes beyond aesthetics and fresh fruit, as it also acts as a giant thermal battery, storing heat in the plants, soil, and concrete surfaces so that it can be used in winter to warm the rest of the house.

Designed by Amory Lovins, scientist, energy efficiency expert, and cofounder of the nonprofit Rocky Mountain Institute (now just RMI), the house has been both his family home and a giant science experiment.

Lovins envisioned an affordable, pleasant house that essentially rejected multiple basic design principles that have governed most of our buildings for centuries. He believed it was unconscionable to still construct walls and windows that behave like Swiss cheese when it comes to retaining heat in winter and keeping it out in summer. That design flaw was only compounded by an equally wasteful solution: using inefficient fossil fuel heating and electricity-hogging air-conditioning, both of which have to work overtime at great cost to budgets, climate, and environment to counter the Swiss cheese problem, which he maintained was insane on its face. Heating and cooling a poorly insulated house was like bailing out a boat with a hole in the hull. You could keep afloat that way, but it's exhausting. Such design choices might have been the norm in the past, Lovins said, but we can do so much better now. The Banana Farm would be his proving ground for his belief that buildings, energy systems, transportation, and basically every major component of modern human life could be radically more energy efficient, sustainable, and resilient, and thereby far less harmful to planet and climate. His selling point, though, would be that his take on what should be standard building design and technology made more sense economically than the conventional way of doing things—it would not only pay for itself; it would add wealth.

In short, Lovins wanted to replace building designs that combat nature using the brute force of fossil fuels with those that use what nature offers for free in the form of sun, shade, and the fundamental laws of thermodynamics. He wanted to spark a revolution that would change the way we build forever.

"Renewables get virtually all the headlines because they're visible, while energy is invisible," Lovins says. "So the energy you *don't* use is almost unimaginable. Energy efficiency . . . is hard, but rewarded by savings worth trillions of dollars."

To say Lovins's ideas were greeted with skepticism would be an understatement. He had, after all, branded the design practices used by the entire construction industry as "stupid."

So the stakes were pretty high when he broke ground on the place, an unconventional home of curved walls and massive windows. Anything could be made to work if you threw enough money at it. The question he had to answer was: Could he really build a home that not only worked but that could be widely replicated at a reasonable price? Or would the Banana Farm turn out to be nothing more than a wishful-thinking demonstration of someday possibilities, no more practical than Edison's City of Light was in 1893, or the cartoon flying car from *The Jetsons* still is?

The answer, according to Lovins, is that it's not only possible to have an affordable, passively heated and cooled, low-waste house that also can be energy self-sufficient in times of blackouts or disasters; it's absolutely essential—for the future of our economy, health, climate, environment, and energy security.

As for practicality, there is one more thing about the Banana Farm to keep in mind, its most remarkable quality: Lovins built it in 1984.

That's right, Lovins accomplished something conventional wisdom said was impossible four decades ago. He did all that using early 1980s technology—tech that predated not just smartphones but all mobile phones smaller than a briefcase. He did this before there was a World Wide Web, when the Commodore 64 (basically a souped-up keyboard that could be connected to your cathode-ray tube TV set) was the most popular home computer, and when computer viruses were only a theoretical problem. He has since upgraded the house with better solar panels, better appliances, and a battery bank to store his solar power. Everything Lovins needed to accomplish the efficiencies in the Banana Farm has gotten better and cheaper in constant dollars. His first solar electric panels were less than half as efficient as today's rooftop products. He paid $26 per watt in 2023 dollars ($8.77 back then) for the solar energy capacity he purchased. Today's panels cost *twenty-six cents* per watt of electricity (nine cents back in 1984). The Banana Farm model is better and more doable now than ever before.

While the central United States was pummeled with an arctic blast that left 4.5 million Texans shivering in frigid, dark homes during 2021's deadly winter blackout crisis, triggered by frozen natural gas–fueled power stations ill-equipped for a new age of extreme weather, the Banana Farm ticked along without a hiccup. Even in the harsher conditions and lower temperatures a thousand miles to the north, it needed no furnace other than the sun.

If they had incorporated just a fraction of these design principles, Texas homes and businesses would have been far less vulnerable to power outages and extreme weather, perhaps blunting the blackout crisis's brutal death toll—246 men, women, and children, according to the official count, though some independent analyses put the deaths from the weather and blackouts from 700 to more than 800. With the Banana Farm's insulation and windows, a home can stay warm just from the body heat of its occupants, or from the oven being on long enough to bake a batch of cookies, no atrium required. The 100 million Americans living under official heat-wave warnings during the fiery summer of 2023—nearly a third of the US population—could have remained cool and safe with just some of the Banana Farm's features without overloading the grid from massive air-conditioner use. Nationwide, billions of dollars in energy costs could be avoided, reducing the carbon footprint of our buildings while making our communities far more resilient in the face of extreme storms and temperatures.

The Banana Farm, now a sustainability touchstone for more than a hundred thousand visitors, *did* launch a revolution. It's just taking a lot longer to kick in than Lovins had hoped. The word is spreading now, though. Lovins and RMI moved from being outliers in the 1980s to energy and design influencers in the new century, hired by such companies as Ford, BMW, Coca-Cola, Walmart, Bank of America, General Motors, and more than one hundred energy utilities, along with the United Nations, thirteen states, and the US Congress, Department of Energy, and Department of Defense. He's advised government and business leaders in more than fifty countries and twenty-three heads of state on energy efficiency. Lovins and RMI have applied the principles of the Banana Farm to other construction, both new and through upgrades of older buildings, including some iconic structures. They led a deep retrofit of the Empire State Building in 2012 that including new insulated windows and efficient lighting, which allowed them to shrink the cooling system. Energy use was reduced by 40 percent, which meant $4.4 million a year saved in utility bills. The retrofit paid for itself in three years. A subsequent retrofit of Denver's old government complex was even more dramatic, cutting the energy bill by 70 percent. The Banana Farm principles pay for themselves, and since 120 million US buildings combined consume more electricity than every country that's not the United States or China, that's a very big deal. In addition to making homes and communities more economical and resilient in an age of increasingly extreme weather, it could put a serious dent in the country's annual collective half-trillion-dollar electric bill.

From Banana Farm to Passive House

The Banana Farm has also helped inspire the next generation of building sustainability, a movement called Passive House. This is a voluntary design standard and certification for super-energy-efficient, superinsulated buildings that stay warm and cool without furnaces or air conditioners. It has flourished in Germany since the 1990s (the original name was *Passivhaus*), then spread throughout Europe. Now, all these years later, it's gaining momentum in North America. In many ways, this is the revolution Lovins predicted because it takes his concept mainstream.

Unlike the Banana Farm's distinctive curves and its central living atrium, Passive House buildings can and usually do look like conventional homes, apartment buildings, schools, and offices. Any type of house can be made Passive House compliant, but the age-old boxy colonial design with a steeply sloped, peaked roof—basically the box with the triangle on top that nearly every child produces the first time they draw a house—is ideal in its simplicity and low cost. Since an essential feature of the Passive House design is airtightness when windows and doors are closed, the less complex a structure is, with the fewest bends, nooks, and crannies—where leaks and seals can be difficult—the better.

Besides the airtightness requirement, the key difference from Lovins's home is that modern Passive House buildings don't rely on atriums as heat sources—heat pumps take their place to maintain a constant temperature throughout the house. Solar photovoltaic power is not required, though it is a great plus for the design and for the ability to retain power during electrical outages (and those steep rooftops, with one of them south-facing, are ideal for solar).

The Banana Farm's key features, the insulated walls and windows and the use of passive solar heating, remain vital parts of the Passive House design standards. Next-gen versions of Lovins's first super windows are now much more common, and many Passive Houses favor large south-facing windows as major sources of heat. Thermal energy doesn't need to be stored in an atrium—the airtight Passive House functions like a giant thermos to contain the heat, losing no more than half a degree a day when it is dark and cold, which is what the heat pump takes care of. Unlike a furnace or an air conditioner in a drafty or poorly insulated house, always working hard to play catch-up, the Passive House's heat pump has very little to do to keep a constant climate. That's why the combination of insulation and heat pump is such a powerful one-two push against waste and cost.

But to be livable, this building-size thermos needs a ventilation system, in place of drafts and open windows and attic vents. Of course, you can open the windows anytime you want, and many Passive Houses open up to outdoor living spaces in nice weather, but the default is airtight in cold or hot weather. This turns out to be less of a liability and more of a killer feature: the design provides healthy indoor air quality in an era of extreme weather and wildfires that are increasingly bedeviling whole swaths of the country with choking and harmful smoke and pollution. Passive House design standards require top-flight filtration to remove pollutants, bringing in a constant supply of fresh, clean air and cycling older air back out. The heat pump captures warmth from the outgoing air, keeping it inside in winter and sending it outside in summer, while running all the incoming air through a HEPA filter. This system means the place where families spend most of their time—the home—becomes the one place where they breathe air clean and free of pollutants no matter what's going on outside. Indoor air quality typically is three or four times worse in conventional houses than outside. In a Passive House, most indoor pollution is banished.

High-efficiency appliances and lighting not only save electricity; they eliminate the excess heat that ordinary appliances and light bulbs produce, enabling the house to do the least work possible to stay at a constant temperature in every room. A Passive House succeeds because it is merciless in eliminating waste: it uses up to 90 percent less energy than a conventional home for heating and cooling, and 70 percent less energy overall. Such buildings not only radically cut heat-trapping emissions and other fossil fuel waste, but they also pay for themselves every month. If a household spends around $2,000 for energy every year (that's about average in the United States, with considerable geographic variation), a Passive House family would pay only $600—before any savings they get if they install solar photovoltaics on the roof. That's just $50 a month.

A couple with two young children moved from Philadelphia to the small town of Hope in midcoastal Maine and bought a new home built to the Passive House standard. They added rooftop solar to their 1,500-square-foot two-story house, built in the New England farmhouse style, which knocked down their energy bill to $13 a month. That's the base fee just to be connected to the grid, and because they make four times the electricity they need, they never pay more than that minimum. The fact that's all they pay to keep their place at 70 degrees on subfreezing winter days gives the couple a smile every time the bill arrives—as does the security of having a home that is as resilient as possible in the face of an uncertain climate and unpredictable energy prices.

"We wanted it to be a sustainable design and materials," Madeleine Mackell told Maine Public Radio. "And the fact that we only have to pay thirteen dollars a month for our connection fee is just an added bonus."

This turned out to be more of a bonus than they'd ever anticipated. Their first full year in Maine marked a big spike in the price of heating oil, which 80 percent of Mainers use for home heat in the winter. The average Mainer's combined heating and electricity bills that year ran about $6,000, averaging out to about $500 a month. It doesn't take many years of saving nearly $6,000 in annual energy bills to pay back the added onetime costs of those superinsulated windows and walls.

Single-family Passive Houses can cost 5 to 10 percent more per square foot than comparable conventional homes, but as with the Banana Farm, payback can be fairly quick. Though no matter how long it takes, you're just trading a little extra a month for a mortgage that will eventually be paid off, while shedding energy bills that will not only never go away but also inevitably rise over time. Not having to pay such whopping energy costs actually lowers your monthly obligations substantially in a lender's eyes, which allows buyers of Passive Houses to quality for higher mortgage amounts.

Architects are finding ways to lower Passive House costs to match standard building rates in some parts of the country, or to at least come close. Pennsylvania architect Richard Pedranti has shifted his practice to specialize in this efficient design, and he has tried to lower costs by prefabricating most of a Passive House, building the insulated walls with wiring, plumbing, and ductwork in place at the factory. Then the finished pieces just need to be assembled at the homesite, saving time, labor, and money.

These homes have simple, elegant designs and they are both simple to operate—it's not like living in a spaceship—and inexpensive to maintain. As enticing as single-family Passive Houses may be, the design standard's really big wins for cutting waste and saving money are most obvious with big buildings: multifamily apartments, condo and townhome complexes, offices, schools, and dormitories. The extra costs for insulation and ventilation become minor with large-scale projects, and are dwarfed by the gains from cutting energy waste. That's why the Empire State Building retrofit was such a no-brainer, and why European countries—Germany, Britain, and the Netherlands, in particular—are all using the Passive House design to build affordable apartment buildings and subsidized multifamily housing for low-income families. The energy savings pays for much of the cost of Passive House retrofits of existing subsidized housing as well as underwriting new construction—even with upgraded all-electric kitchens. This is a model for solving affordable-housing shortages and homelessness in cities around the world, and could offer hope to those struggling in the states with the most unhoused families—California, New York, Texas, and Florida. The Passive House design could be a pathway for better living for everyone, including providing healthy, cost-effective, and beautiful housing for lower-income families. And in the process, we would be removing the most wasteful and polluting aspects of our built environment.

The federal government can provide financial incentives to advance this—generous funding, tax breaks, the same sorts of rebates that are advancing electric cars—but what would really drive it forward is communities embracing and favoring new construction that features some, if not all, Passive House design ideas. The threat of extreme weather creating dangerous air quality coupled with the strain on the grid caused by conventional air-conditioning demand during heat waves make Passive House designs not just desirable but essential survival and resilience tools. A Passive House can remain comfortable and safe through long power outages in hot or cold weather—and indefinitely if it has solar power with storage. No conventional house can make that claim.

Selling this positive vision to community leaders and the public could prove to be more productive than fighting culture wars over gas bans, and it would lead to the same outcome. Passive Houses don't need or use fossil fuels.

Meanwhile, universities worldwide are embracing Passive House designs in labs, living spaces, and lecture halls to lure a new generation of students with a high commitment to sustainability and environmental good. Wheaton College has opened a 178-student Passive House dormitory, the first in Massachusetts. The University of Southern Maine is building a new 580-student dorm on its campus near downtown Portland, with a mix of single rooms, studio apartments, and larger apartments with multiple bedrooms and bathrooms, making it the second-largest Passive House university building in the country. The largest is at Cornell University's Cornell Tech graduate school on Manhattan's Roosevelt Island—a twenty-six-story apartment building, billed as the world's first Passive House high-rise. The University of Northern British Columbia opened the first Passive House research lab in North America in downtown Prince George, where the eye-catching gleaming wood structure is home to a graduate program focused on the next-generation sustainable wood design and construction methods used in creating the lab.

All of them are reaping huge savings in energy costs for their institutions and occupants, while radically reducing wasteful carbon emissions and other pollutants normally produced by such massive buildings.

The creator of the groundbreaking Banana Farm is in his seventies now, relinquishing his post as chief scientist at RMI but still an emeritus leader there, and still promoting his vision of energy efficiency and removing waste from every part of our energy system. Amory Lovins summed up these thoughts in a recent speech:

> For us, the first self-endangered species, I have bad news and good news. Conservative climate models underplayed climate change's speed and runaway feedbacks—but they also understated practical and profitable ways to prevent it. Offsetting these two factors, the race of and for our lives is very much on. With neither complacency nor despair, with both sharp impatience and relentless patience, we must double down on what makes sense and makes money.

Fixing the waste in our buildings at the end of the energy chain is vital, Lovins says. It removes a major source of cost, pollution, and climate disruption, and all that added efficiency lowers demand and pressure on our aging and overloaded grid. But we have to fix the front end of the system, too, where even more waste lies—with its battalions of spinning generators and utility power plants crackling with high voltage. That's where the primitive nineteenth-century technology of spark and fire still reigns supreme, machines belching smoke and carbon that even Thomas Edison derided as mere stopgaps until we figured out how to harness solar energy. That's where the beating, bleeding heart of the grid still lives, along with the seemingly impossible truth about our energy—not just our electrical generation, but all the other things we do with spark and fire to heat or drive or ship or farm or fly or manufacture:

We waste two-thirds of the energy we consume.

That, to use Lovins's terms, is where the ultimate battle to make sense and make money lies. Renewable energy isn't just about being green or fighting climate change. It's about saving incredible sums of money and eliminating the greatest source of waste in the human world: burning fossil fuels. Sense and money, indeed.

Without that colossal waste there would be no climate crisis of the magnitude we now face, no increasingly deadly heat spells, no abnormal onslaughts of extreme weather. Our national security and democratic values would no longer be at risk through fossil fuel dependence. And we would no longer have to watch two-thirds of our annual energy and fuel spending—a trillion dollars—go up in smoke, wasted by an energy system that is total garbage by design.

As with our buildings, the waste at the front end of the energy system can only be fixed by replacing spark and fire with renewable power and green technology. The question is: Can these twin energy *Titanic*s, buildings and power plants, be turned in time?

The answer may already be on display in two of the most radically different places America has to offer: New York City and rural Iowa. Call it Broadway meets the Corn Belt, where the future of energy and the end of waste just might be unfolding right now.

Chutes and Ladders

G rowing up in Brooklyn's Bedford-Stuyvesant neighborhood, Donnel Baird remembers how his family kept their apartment warm when the thermometer dropped and layers of sweaters were not enough. The building's ancient oil-burning furnace hadn't worked in years, so they would crank up the gas stove and leave the oven open to provide some heat—a dangerous choice, made tolerable by cracking a window enough to ward off carbon monoxide poisoning. It was the best his parents could do at the time. They were recent immigrants from Guyana struggling to make ends meet but intent on making sure their son received the best possible education, heat or no heat.

That experience stuck with him. After prep school, college, graduate school, and an MBA, he recalls he had two things he wanted to tackle. Working to save the environment was one, and the other was even more personal: "I wanted to give something back to the other kids I grew up with who didn't have my educational opportunities."

No one could ever accuse Donnel Baird of not keeping his promises— or of being insufficiently ambitious about doing so.

Here, in a nutshell, is the business plan he came up with, in the final semester of his graduate studies at Columbia University, to combine the two goals: He wanted to decarbonize America's buildings and help turn our grid green with efficient electric heating, cooling, and solar energy panels for those buildings. *All* of America's buildings.

He would do this city by city, beginning with New York and starting with the sorts of buildings he grew up in: the oldest, least efficient buildings in the most underserved neighborhoods and communities of color.

Finally, he would do this by training a vast Civilian Climate Corps, employing vulnerable populations from places like his old Bed-Stuy neighborhood, including the unemployed and underemployed, at-risk youths, and recently released probationers and parolees. He would build on experience from one of his first jobs out of college, working for a community-based program funded by the Department of Energy to create green energy jobs during the Obama administration.

When Baird launched his public benefit corporation, BlocPower, potential investors and venture capitalists he pitched had their doubts about the young entrepreneur's bold mission to, in essence, rebuild America one city at a time. He explained that his approach would be to "turn America's buildings into Teslas," stripping out the fossil fuel components and replacing them with clean, green technology—while structuring the endeavor as a lucrative, investible business. His would-be backers, particularly the Silicon Valley crowd, loved the sound of that, but they would invariably balk at starting with economically challenged communities. It was clear that they had doubts about the financial model and wished Baird would court neighborhoods that had plenty of actual Teslas tooling around. Gaining traction for his plan was, to say the least, a struggle for Baird . . . at first.

So he began by scraping for funds and contracts, starting small: a community center retrofit here, a church project there, a few modest grants, improving neighborhoods a bit at a time with the magic of heat, rooftop solar, and green jobs. Then the city of New York hired BlocPower to train underemployed New York City residents for green economy jobs. The contract was worth $37 million, more than sufficient to launch a big part of Baird's dream: his Civilian Climate Corps, with enough to train a thousand men and women.

As a handful of building retrofits in New York turned into dozens, then hundreds, skepticism eased. Fast forward ten years to the present:

The company is expanding nationwide. It's under contract to decarbonize all of Ithaca, New York's 6,000 buildings by the year 2030, and then all 10,000 buildings in Menlo Park, California. More large-scale electrification and renewable energy projects are on tap for Denver, Oakland, and San Jose. The dream of electrifying private, public, and government buildings city by city has begun to seem less pie-in-the-sky.

By 2023, BlocPower had raised more than $250 million in capital investments and debt financing from such investors as Goldman Sachs, Microsoft's Climate Innovation Fund, Van Jones, NBA great Russell Westbrook, and others. These investments enabled Baird to launch a financing plan to pay for energy retrofits and community solar projects in financially stressed cities, towns, and neighborhoods. Just as Amory Lovins and RMI demonstrated when retrofitting the Empire State Building—that green and efficient energy systems can pay for themselves—BlocPower's business model leverages monthly energy bill savings to finance heat pumps, efficiency upgrades, and renewables in old apartment buildings, churches, community centers, and other buildings. The company's goal with each retrofit is to cut utility bills by half or more.

This is the prospect that persuades building owners to hire BlocPower. The barrier has always been the upfront equipment and installation costs, often prohibitive for building owners in underserved communities. So BlocPower tries to overcome this age-old problem of access and opportunity by doing the retrofits through a no-money-down, monthly lease-to-own program. When all goes to plan, some of the utility savings stay with the building owners, with most going to BlocPower to pay for the energy upgrades over time—a fifteen-year financing plan for green tech similar to a home mortgage. In theory, everybody wins—especially the environment.

In one case study shared by the company, energy bills declined dramatically from $32,178 to $14,506—a savings of $17,672, more than enough to cover the retrofit over time. But BlocPower does not guarantee such results, and there have been reports of more meager savings or outright energy cost increases in some old and problematic buildings.

Although Baird says this approach achieves energy cost savings in at least nine retrofits out of ten, the unpredictable and sometimes disappointing results show just how challenging it will be to scale up mass decarbonization of the nation's aging building stock.

Still, BlocPower's model has drawn headlines, accolades, and imitators. The White House named Baird a "Champion of Change" for this work, while *Time* magazine placed BlocPower on its "Time100 Most Influential Companies" list for two years running, additionally naming Baird its first "Dreamer of the Year" in 2022. The New York–based green developer Sealed, meanwhile, is partnering with utility companies to offer similarly financed green upgrades to homeowners.

Several hundred million in government grants and contracts have also poured in, including $108 million more from New York City for BlocPower to train three thousand residents with the Civilian Climate Corps. Baird says the program is modeled on President Franklin Delano Roosevelt's Depression-era jobs and infrastructure program, the Civilian Conservation Corps—a 3-million-worker green army that planted 3.5 billion trees, built 711 state parks, created thousands of public campgrounds and recreational sites still in use, and protected more than 40 million acres of farmland from erosion and ruin during the infamous Dust Bowl era in the Plains States.

"President Roosevelt's Conservation Corps in the Great Depression trained and hired able-bodied people to . . . lay the foundation for overhauling the infrastructure of America," Baird says. "Well, we want to lay the foundation of sustainability across America, decarbonizing buildings and decarbonizing whole cities. . . . The dream of BlocPower is that we can go into an undervalued community and train and hire people to transform their own community and lead the way on helping all of America learn how to solve the climate crisis."

The remaking of buildings, communities, and the machines that power, heat, and cool them is an essential step in solving the colossal waste embedded in our energy system. This need not be an expense with no return; it can be an investment in infrastructure and jobs that pays financial *and* environmental dividends. The same cannot be said for the current fossil fuel–driven system, which wastes the bulk of the trillion dollars we spend on energy every year and always will—until it's replaced at last.

It's Not the Power We Use

The economic case for swapping our fossil fuel–dominated energy system for all renewables is simple: The problem is not just the dirty power we use. The bigger threat is the power we *waste*.

But how, you might ask, do we know that? How do we know that two-thirds of our energy is wasted every year?

It's no secret—the data is public, published on the web for all to see. It simply doesn't get the attention it deserves, which is pretty crazy, given that most people would agree that two-thirds of a trillion dollars a year is a mind-boggling amount to waste.

And yet, every year, the Lawrence Livermore National Laboratory dutifully labors in obscurity, tracking and measuring how much power we consume from all sources and methods, how much we generate, how much we use productively, and how much goes down the drain. That last is the energy we pay for, but waste—by far the largest part of our energy system.

Livermore turns all that data into a compelling, strange chart, a fuel-soaked take on the venerable Chutes and Ladders game, though this pictorial representation of our many rivers of energy is known as a Sankey diagram (see next page).

The numbers are so big that this annual chart uses a measure called a quad as a kind of shorthand, with "quad" defined as 1 quadrillion BTUs of energy—that's a 1 followed by 15 zeros, a very big number. You'd need 125,000 identical planet earths to have a quadrillion people. So, *really* big.

A BTU, or British thermal unit, on the other hand, is very small. It is an old standard measure: the amount of energy needed to heat one pound of water 1 degree Fahrenheit. One typical gas stove burner puts out 7,000 BTUs of heat energy an hour. One quad equals the amount of energy

Estimated U.S. Energy Consumption in 2021: 97.3 Quads

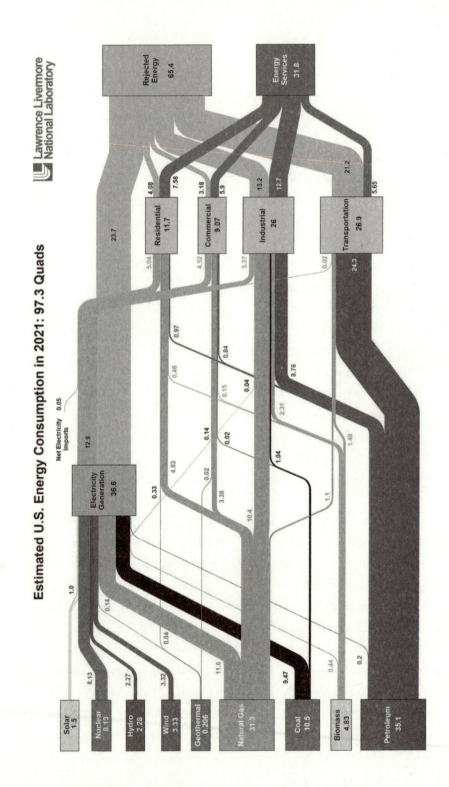

Lawrence Livermore National Laboratory

released by burning about 55 million tons of coal. The atomic bomb that devastated Hiroshima released 5.7 quads of energy.

Last year the United States consumed 97.3 quads of energy, most of it from burning oil, gas, and coal, followed by nuclear power and renewable energy—wind, solar, hydropower, and geothermal. The left side of the chart shows all those inputs flowing to the right into transportation, to generate electricity—the single largest quad-eater—and for residential, commercial, and industrial uses.

On the chart's right side, everything flows into two end points. They are not equal. The "energy services" end gets 31.8 quads—that's the energy that actually moves your car or heats your home or powers all the machines that run on electricity. The larger end point, with 65.4 quads of costly power—67 percent of the total energy supply—is "rejected energy." Better known as waste.

And this is a *lot* of waste. With a little more than 4 percent of the world's population, Americans consume nearly 20 percent of the world's energy output. So our waste isn't just a national disaster—it's a global tragedy. Person for person, we waste more energy than most countries

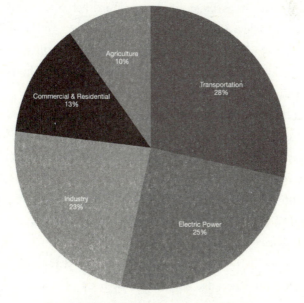

Sources of US greenhouse gas emissions, 2021, EPA

produce—a list that includes Spain, Italy, the United Kingdom, Germany, and Canada.

The biggest energy waste culprit is riding that thick, dark "Petroleum" chute at the bottom of the Livermore chart: transportation. Just under 27 quads go in, but 21.2 of them are wasted. That means about 80 percent of our transportation fuel lands in the metaphorical garbage bin, bled off through heat loss and friction, making it our leading climate disrupter. This is due to the massive inefficiencies that are built into the internal combustion engine, which from a physics point of view is primarily a heater that uses a fraction of its energy to move the car. In comparison, electric cars are 90 percent efficient because they don't set fire to anything or transfer the resulting heat into mechanical energy. The battery power simply goes directly to the motors that turn the wheels through efficient magnetic induction. If that sounds familiar, it is—it's the same reason that electric heat pumps are more efficient than fossil fuel furnaces and that induction stoves outstrip gas burners. It's basic and unalterable physics.

The Livermore chart shows the generation of electricity as the single largest consumer of quads, and its wastefulness mirrors the entire energy system's—two-thirds garbage, while one-third actually makes the lights and motors and television sets and surgical theaters and a million other things work. Because that's not as terrible as transportation's waste, the power-generation sector manages a slightly lower greenhouse gas footprint even though it uses more energy.

One hundred and thirty years ago, spark and fire and fossil fuels were essential, providing the energy we needed the only way we knew how. They nurtured our industries, protected us in war, lit our homes, helped put food on our tables, and made possible much that is good as well as bad in modern life. We knew spark and fire and fossil fuels were poison even then, though not the extent of the damage they held in store for health, wealth, and planet. It was the best we could do—then.

But it stopped being the best we could do fifty years ago. Now it's the worst thing we do—because now we know better. Because now we have options. Because now the waste is a choice, not a necessity.

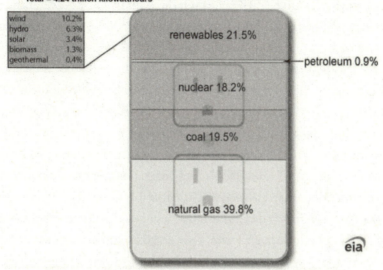

Sources of U.S. electricity generation, 2022
Total = 4.24 trillion kilowatthours

wind	10.2%
hydro	6.3%
solar	3.4%
biomass	1.3%
geothermal	0.4%

renewables 21.5%

petroleum 0.9%

nuclear 18.2%

coal 19.5%

natural gas 39.8%

eia

Source: US Energy Information Administration

Renewables for All

The good news on that Chutes and Ladders diagram is that 40 percent of the electricity supply looks and acts differently, because it's carbon-free. Roughly half of the slice of that carbon-free pie is created from renewable sources—wind, solar, hydropower, geothermal, and biomass energy, with rapid growth in solar and wind currently underway. The rest of the carbon-free energy comes from nuclear power plants. Twenty years ago, when more than 70 percent of electricity came from fossil fuels, the amount from renewable energy was basically a rounding error, except for hydropower plants. It's not that renewables are perfectly efficient—solar panels convert only some of the sunlight striking them into electricity, and wind turbines cannot capture all the wind's energy. But no "fuel" is wasted, and there are no emissions or waste heat, so any efficiency comparisons with fossil fuel machines are meaningless. Efficiency for renewable energy isn't about energy waste; it's about the dollar cost of solar panels or wind turbines per watt of electricity they generate.

Renewables are growing because the economics that once made them too expensive for mass usage by a utility—remember the $26 per watt Amory Lovins had to pay when he built the Banana Farm?—have shifted dramatically. Those costs have decreased a hundredfold, and the solar and wind technology has gotten so efficient so fast that renewable electricity is now the cheapest energy there is. Setting aside the pollution, health, and climate benefits, the economic case for renewables isn't even close: In 2022, electricity produced from solar energy cost 33 percent less than power generated by burning natural gas. Wind electricity was 44 percent cheaper. Electricity from burning coal was more than twice the price of its renewable competition. And unlike fossil fuel generation, renewable power can be local and home-based as well as built at utility scale. The combination of rooftop solar, farm- and campus-based wind, and renewable utility scale installations of both is a zero-waste, zero-carbon solution that can't be beat. Now there is a new player coming online: offshore wind projects, which tap into nearly limitless marine wind resources. This is the next big thing in the US and is already being deployed elsewhere in the world, with Denmark leading the way. Together, the combination of local and utility-scale renewables has rendered fossil fuel power uncompetitive in every way.

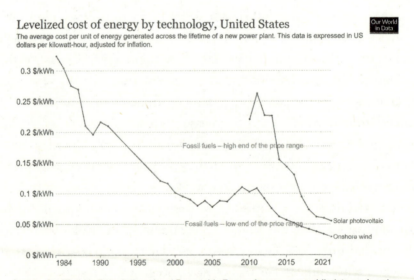

Levelized cost of energy by technology, United States

The average cost per unit of energy generated across the lifetime of a new power plant. This data is expressed in US dollars per kilowatt-hour, adjusted for inflation.

Our World in Data

Source: Our World in Data, International Renewable Energy Agency, ourworldindata.org/grapher /levelized-cost-of-energy. Note: Data expressed in constant 2021 US dollars

This is why many states and whole countries elsewhere in the world have set ambitious goals to decarbonize their electricity through solar and wind expansion, with California required by state law to have a green grid by 2035 and only electric cars for sale by 2030. The economics are so good for renewables now that a truly astonishing number of utility-scale, investor-backed projects are ready to roll: 2,000 gigawatts of solar, wind, and energy storage proposed all over America. The number is astonishing because the nation currently generates only 1,200 gigawatts from all energy sources combined (and only 300 of those gigawatts are renewable). So this would be the upgrade of all upgrades, enough juice to electrify everything renewably twice over.

This renewable investment avalanche shows where the money and opportunity are at now in the energy business, but there is a downside. This unprecedented number of projects has created a backlog in the approval process, with a delay of four years or more. Part of the slowness is dysfunctional bureaucracy, but a lot of it is due to the limitations of the electrical grid itself—the three regional networks of wires, transformers, and transmission towers that carry electricity around the country. America's grid is a bit of a laughingstock to the rest of the developed world: aging, inefficient, fragmented into three separately governed and rarely cooperative entities, increasingly subject to blackouts, and long overdue for an upgrade. Most of this vital infrastructure hasn't been new since the 1990s or before; parts of the grid date back to the Kennedy administration. It simply could not handle all the proposed new renewable power if all the projects were quickly approved, nor could it efficiently move all that green electricity from places with large solar and wind resources to areas that lack them. In 2022 the federal government launched a $20 billion grid modernization program to begin tackling this critical problem—which hampers the movement and reliability of all forms of electricity, not just renewables.

Even with massive investment in renewable power and the grid, switching to a carbon-free and nonpolluting power system is a monumental task. It requires pushing out the most entrenched and powerful industrial incumbent in history, along with all the investments and jobs tied up in the fossil fuel business. But it's by no means impossible. The average

American power plant is thirty years old and not getting any younger, so it's not like we'd be tossing out something new and shiny. We have a long history of tossing aside similarly entrenched incumbents, things for which we also had habitual and even sentimental attachment. We replaced horses with cars early in the twentieth century in a relative blink of an eye, killing millions of jobs and an entire economic ecosystem of breeders, ranchers, trainers, dealers, stables, saddle makers, blacksmiths, buggy and wagon manufacturers, feed growers, and veterinarians. Building the interstate highway system upended the freight industry, as well as tourism and trade for a host of cities that became ghost towns in the name of progress. Cellphones devastated landline companies. The film camera's century of utter dominance met an unceremonious end via digital cameras, which soon died their own quick death by smartphone. And factory farms crushed family farms, once the backbone of heartland America. Nearly a third of Americans lived on farms in 1920, but less than 2 percent call them home today. For better and sometimes for worse, we have a long history of replacing billion-dollar industries and entire ways of life with astonishing speed when it suits us. At one time or another, each of those shifts was thought to be impossible. And then one day, they weren't.

The fossil fuel industry has dreaded and fought off the inflection point we have finally reached on energy, where the impossible gives way to the possible. This is why there has been a proliferation of myths and misinformation about renewables in recent years: claims that windmills are noise-polluting health hazards, solar farms are killing endangered species and using up too much valuable land, and renewables can't be counted on because they are "intermittent," while only fossil fuel power plants can ensure "reliability."

None of this is true. It's always sunny or windy somewhere, so the more solar and wind farms we build, the less intermittency can be an issue. Carbon-free nuclear power, hydropower, and geothermal energy plants *are* capable of running 24-7—and currently supply more than 25 percent of the nation's electricity, which is more than enough to fill any gaps on those rare cloudy, windless days. The rapid growth in utility-scale battery storage plants now underway can also make solar power available at night and wind power when there is no wind. As for using up land: building enough renewable energy to power the whole country would require less than the 1.3 percent of total US acreage currently occupied by fossil fuel infrastructure. The claims of harm to nature and wildlife are particularly disingenuous, as any impact on nature caused by renewable energy construction pales in comparison to the daily species-killing, ocean-warming, forest-destroying toxic emissions and climate disruption driven by fossil fuels. It's like complaining to the landlord about a leaky faucet while floodwaters are about to drown you. As for reliability, it was the fossil fuel power plants, not renewable energy installations, that failed during those deadly winter storms in Texas in 2021. Renewable energy is resilient. The homes with rooftop solar and battery storage will still have power when the lights go out and the heat stops working everywhere else.

There is even a renewable energy solution for the 50 percent of Americans who can't choose a solar-powered home, either because they are renters or because their roof or its exposure to sunlight isn't suitable. It's called community solar, sometimes referred to as "solar gardens" because they work like community vegetable gardens. Except the crop grown in these gardens is clean electricity shared with neighbors through subscriptions. A typical example is the community solar installation built by BlocPower on the rooftops of a Staten Island church and the community center next door, which together supply the surrounding community with clean power. In Minneapolis, the Shiloh Temple International Ministries recently hired a solar power company to install a large 630-panel community solar garden on the Pentecostal church's roof.

Nationwide, community solar gardens generate 3,200 megawatts of power as of 2021, which is equivalent to the electricity produced by one and a half Hoover Dams, enough to power 3 million homes. Florida is the leader in community solar with half of those 3,200 megawatts, followed by New York, Minnesota, and Massachusetts. Such projects also build community resilience, health, and safety in this age of harsh weather and wildfire smoke. With new federal grants available and twenty-two states putting new laws in place, community solar is growing as a fairly painless way for anyone to benefit from solar energy. Some cities, town, states, and even local homeowners' associations have been slow to embrace the concept, however. This is where community advocates—Sarah Nichols's Marges—can make a difference by working to clear any local indifference or obstacles for renewable energy, efficiency, and electrification. Building codes and policies are mostly a local thing, so communities have tremendous leeway for crafting rules that can either welcome renewable energy projects or make them dauntingly difficult. Tell your local representatives you want the welcome wagon, not a roadblock, so that any group with open space or a large, warehouse-size rooftop —houses of worship, schools, towns, civic groups, condo associations—can create a solar garden. Or you can subscribe to an existing community solar installation in your area— there are web search tools to help with that (see the "What You Can Do" section at the end of this chapter).

The only real obstacle to the renewable shift is the same one we face with every other form of embedded waste: habit. We are used to what we are used to. That is always the barrier to change, the inertia and the doubts—right up until a tipping point is reached, when the new's advantages over the old are simply too great to be dismissed. And we have reached that tipping point on renewables. Between the declining costs and the new federal incentives for embracing them, from utility-scale megaprojects down to the rooftop panels on our homes, we are living in a renewable moment. Why else would it be the red states of Texas and Iowa that lead the country on wind energy?

Bolstered by the new economics of wind, Iowa is leading its own renewable revolution, every bit as game-changing as BlocPower's New York–based infrastructure makeover, remaking and greening the grid down on the farm. One engine driving this transformation of the grid in Iowa is the rural electric cooperative, a tool that empowers farmland Marges to build locally owned and managed wind and solar projects in heartland America. In Iowa, farmers are embracing renewables not just as an extra or an outlier, but as a lifeline and a livelihood.

The Iowa Miracle

By late October, the gently rolling farmland of central Iowa is stripped bare, its fertile miles of cornfields shorn by harvesters and swept clean by the incessant prairie wind. "We just finished the harvest last night," Randy Caviness said as he gazed fondly at his fields, his weary face as stubbly as his 3,200 acres of corn and soy. "There's only one crop left."

He turned his old Ford pickup onto a narrow Adair County farm road, then braked and pointed. "There it is. We harvest that crop every day, year-round."

A lone wind turbine dominated the landscape, a huge, gleaming, brilliantly white obelisk rising improbably from a drab flatland of browns and grays. It seemed jarringly out of place and proportion, like a skyscraper amid mud huts. From base to blade tip, this 1.6-megawatt windmill stood four hundred feet, taller than the Statue of Liberty's torch. If it were a building, it would be the third highest in the state. The three-bladed fan at its summit spanned an area the size of a football field. Despite a stiff breeze, those massive blades, each as long as three New York City subway cars, seemed to rotate lazily, but Caviness said that's just an optical illusion born of size and perspective—the blade tips reach speeds of 100 miles an hour.

"I'm amazed every time I see it," the lifelong farmer said. "That one turbine powers a whole town. And more."

I first saw Caviness's project more than ten years ago, and he has been riding Iowa's wave of wind power development ever since. The eight turbines of the community wind project he launched have delivered 12.8 megawatts to local communities and rural utilities for all these years, enough juice to power six thousand Iowa homes. They also generate healthy annual returns of up to 16 percent on the $12,000 to $50,000 individual investments from 180 local farmers. And they generate some of the cheapest electricity in the country. Farming the wind next to the other cash crops makes sense and makes money. That's motivation enough for the Corn Belt. Climate doesn't even have to be part of the conversation. This is why Iowa is a national and global leader for farming the wind, its latest and most dramatic crop.

Iowa generates 60 percent of its electricity with wind annually, the best in the nation (on some days it's a lot more than 60 percent). If it were a country, Iowa would be in second place as world wind leader, behind only Scotland, which generates an incredible 97 percent of its electricity from renewables, mostly wind, with some hydroelectric and a bit of solar and waste-to-energy thrown in. Perennial wind technology leader Denmark comes in at a close third, with 55 percent of its power coming from wind.

Caviness says the turbines are a source of local and state pride.

"Everyone loves us. Our turbines make electricity cheaper than they can buy it elsewhere. Obviously, we need to do more of this—in the state and the rest of the country."

WHAT YOU CAN DO TO "SQUEEZE THE JUICE"

There are seven compelling reasons why we can and should do our part to just say no to energy waste by electrifying everything, including our rooftops, right now:

1. Major power blackouts are more frequent and last longer than they did ten years ago. Extreme weather is the primary reason. Rooftop solar offers homeowners a lot less to worry about when outages strike.

2. We now know natural gas is as bad for the environment and climate as burning coal. Unexpectedly large leaks in the system spew the powerful heat-trapping pollutant methane into the atmosphere, shattering natural gas's reputation as the "clean" fossil fuel. All-electric homes and businesses can help make that all go away.

3. Natural gas stoves pump harmful air pollution into our homes and create a 42 percent higher risk that children will experience asthma symptoms. You don't have to replace an entire gas range to minimize this risk. Using a tabletop induction burner and other countertop electrics and keeping the gas range as a backup can bring many of the same benefits on the cheap.

4. Using an electric heat pump with air filtration can cut heating and cooling costs by 50 to 90 percent, depending on what it replaces. The filtration is a safeguard to keep wildfire smoke and other pollutants out of the air we breathe at home. Passive Houses do all that and more, with superinsulation that can almost zero out energy bills.

5. The federal Inflation Reduction Act championed by President Joe Biden offers homeowners and landlords rebates of up to $14,000 per home to cover everything from heat pumps and home insulation to electric stoves and thermally efficient windows and doors. Many states offer additional energy efficiency rebates as well. Not only do these upgrades pay for themselves over time; these rebates lower the buy-in to make the most efficient choices the absolute cheapest choices. The rebates can be up to $8,000 for heat pump installation; $1,750 for a heat pump water heater; $840 for an electric stove; $600 for energy-efficient windows and skylights; $1,600 for insulation, air sealing, and ventilation; $2,500 for electrical wiring improvements; and $4,000 for electric "smart" panel upgrades.

6. Simple efficiency upgrades and practices can save money and cut emissions:
   - Use efficient LED light bulbs.
   - Power down computers at night instead of letting them sleep.
   - Use ceiling fans or window fans to lower the need for air-conditioning.
   - Cool and heat only rooms in use rather than the entire home. Smart systems can be set to do this, or you can just shut the doors to the rooms and the heater vents inside manually.
   - Turn the thermostat down to the low sixties at bedtime in winter.
   - Use shade trees, blinds, or awnings to passively cool the home and reduce the load on air-conditioning in summer.
   - Set your washing machine to cold water, wash full loads, and consider line-drying clothes on nice days.

- Cutting your time in the shower, replacing grass with less-water-consuming plants, relying on the dishwasher instead of handwashing, and using water-saving fixtures and appliances reduce water waste and utility bills, and also save electricity and related emissions from powering community water systems.
- Many appliances—TVs, streaming boxes, cable boxes, game consoles, computers and peripherals, phone chargers not in use, and other "electric" bricks for devices—all suck power when plugged in, whether they are being used or not. Unplugging them one by one is a pain, but plugging them all into a single power strip with an on-off switch can allow these "vampire electronics" to be shut down easily when you sleep or are away. This is a bigger problem than most people realize. A 2022 Natural Resources Defense Council report found that Americans spend $19 billion a year on wasted electricity from vampire devices—an average of $165 a year for each American household, though areas with higher rates push the cost up to $440 a year. Cable boxes and game consoles are the two biggest offenders.

7. You can subscribe to a community solar garden and enjoy the benefits and savings of renewable energy. Find community solar near you by searching these websites:
   - Solar United Neighbors: solarunitedneighbors.org/go-solar/community-solar/find-a-community-solar-project/
   - Energy Sage: energysage.com/shop/community-solar/
   - For an excellent guide to community solar options, see "Half of Americans Can't Install Solar Panels. Here's How They Can Plug into the Sun," by Michael J. Coren, *The Washington Post*'s climate advice columnist: washingtonpost.com/climate-environment/2023/10/10/community-solar-renters-apartments-discounted-electricity.

PART III

Eating Up, Driving Off, Buying Out

Imagine walking out of a grocery store with four bags of groceries, dropping one in the parking lot, and just not bothering to pick it up. That's essentially what we're doing.

—Dana Gunders, ReFED, quoted in the 2014 film *Just Eat It*

8

Stick a Fork in It

A change in diet is a way of saying simply: I have a choice.
That is the first step. For how can we take responsibility for the
future unless we can make choices now that take us, personally, off
the destructive path that has been set for us by our forebears?
—FRANCES MOORE LAPPÉ, *DIET FOR A SMALL PLANET*

As soon as he saw that lush front lawn on the corner of Angeles Vista and Olympiad, Jamiah Hargins knew *this* was the place. The quintessentially Los Angeles stucco home with the red tile roof in the View Park neighborhood was nice enough. But what captivated him was the eye-popping green patch of well-trimmed grass sparkling with morning dew, that time-honored symbol of prosperity, homeowner pride, and humanity's dominion over nature he had been looking for.

He pulled over and stared at the lawn, imagining what it would be like to take control of it, this perfect grassy rectangle on this perfect hilly spot on the perfect LA street in just the kind of aging but still graceful and diverse neighborhood he had long sought. It would be the culmination of a year of planning, saving, and dreaming. And as he sat parked in front of that lawn, all he could think about was how great it would be to rip it all out. Every inch, every blade, every root.

Jamiah Hargins dreams of a city with no lawns. And in their place, he sees farms. Lots and lots of farms.

Hargins gave up a successful career in nonprofits and finance to seize this new opportunity (or pipe dream, as his skeptical former colleagues saw it). Even he wondered sometimes if he was crazy. After all, he wasn't trying to eradicate weeds. Or cockroaches. He wanted to kill an American classic, the lawn, an icon of the human-built world so prevalent that its absence seems inconceivable and its presence beyond question, as enduring a part of our daily lives as rush hours and fast-food drive-throughs and fire hydrants. They just are.

Except to Hargins, our grass lawns may be the most senseless and wasteful inventions in history, a bad habit masquerading as tradition, a status symbol sustained by diabolically clever marketing. How else to explain why generations of Americans see costly, labor-intensive, monochromatic green squares of invasive species as not just normal but also beautiful? If only everyone could see them as he does, Hargins laments, they'd recognize lawns as chemically drenched ecological wastelands that nourish no butterfly, bee, bird, or human—just the voracious appetite of the $126 billion lawn care industry.

"Lawns may be green in color," Hargins said, "but they are not *green*. But where there's waste, there's also opportunity."

Which is why he approached the owner of that California bungalow that so caught his imagination and asked him a simple question. "Why mow your yard when you could be *eating* your yard—and feeding the rest of the neighborhood, too?"

Then he laid out his audacious plan. He would build an advanced, super-productive urban microfarm on that gloriously green, wasteful patch of grass. This microfarm would outshine the big industrial agriculture operations that make everything from the contents of boxed macaroni and cheese to those hard, pale tomatoes sold in supermarkets. His food would beat the big boys every time on nutrition, efficiency, sustainability, water use and, last but not least, deliciousness. And his project would be his beta test, his proof of concept. Then he and his nonprofit, Crop Swap LA, would start killing lawns and planting microfarms throughout the city.

"You'll be providing healthy food for yourself and the community," Hargins said, framing the project both as building resilience and striking a blow against food apartheid. "Everybody is entitled to the nutrients beneath their feet. They have the right to have local food. Folks can have a different life. I'm hoping to show that it can be done."

Hargins has the engineering chops that enabled him to win a NASA Mars colony design contest while still in high school, and he's also a born salesman, using his sunny disposition and dazzling smile to good effect. He really wanted to close this deal. Charming as he is, though, a cold call on the doorstep from a total stranger who wanted to kill his lawn could have easily ended with a door slammed in his face. As it happens, the urban farmer and the man with the lawn, Mychal Creer, were acquainted—the homeowner taught at the same charter school that employed Hargins's wife. So he didn't get a door slam—he got a *Tell me more.*

So he did. With a solar-powered, water-conserving irrigation system and regenerative, organic farming methods, Hargins envisioned a powerhouse microfarm raising bounties of vegetables and fruit. The $35,000 price tag would be covered by a mix of grants and rebates from the city, which pays residents to uproot their water-hogging lawns. By uprooting just that 950 square feet of lawn, he promised, they could provide food for up to fifty families in the neighborhood, who would pay a subscription for weekly boxes of fresh produce in what was officially a Los Angeles food desert. The homeowner would get a share for free, and 10 percent would be donated to community fridges and anti-hunger programs. And the whole operation would use only 8 percent of the water the grass currently sucked up.

Once it was going strong, Crop Swap LA would move on to the next lawn conversion. Then the next. And then . . . Hargins conjured an enticing image of a city filled with front yards that yielded healthy, locally sourced, chemical-free, carbon-neutral food—a much greener Los Angeles lawn than one covered with grass—and that focused especially on bringing that affordable fresh food to neighborhoods where it was difficult or impossible to come by. The cost for a monthly subscription for weekly fresh veggies and fruit: $36. With delivery, $47. Much of the labor would be handled by volunteers, so the operating costs would be minimal.

"Let's do it," Hargins pleaded.

"I'll think about it," the homeowner promised.

This was a wildly unconventional take on the sacrosanct American lawn. Would it be unsightly? Or create traffic congestion? Would it lower property values? Would the neighbors object? There was a lot to worry about. But after mulling it over for a week or so, Creer finally agreed to an edible front yard. Hargins hugged him and said he had already thought of a name for the new growing place: it would be the Asante Microfarm, after the Swahili word for "thank you."

It had been a long and complicated journey for Hargins. As a military child in the 1980s, he had a nomadic life—his culture, landscape, school, and friends overturned and replaced every few years. The one comforting constant that he found waiting for him at every Air Force base where his engineer dad was sent was the American-style grass lawn. The irony would not be lost on him in later life, but at the time, he took comfort in that familiar green expanse, its moist coolness, the hiss and chuckle of the sprinklers turning on.

He earned degrees from the University of Chicago and the Columbia University School of International and Public Affairs, dropping the early interest in engineering that had won him that NASA prize and instead choosing a career focused on social justice and racial equity reform. After school he worked for community service nonprofits in the United States and Brazil (he's fluent in Portuguese and Spanish), then became a financial consultant to several global nonprofits. Then in 2015, he and his new wife decided to settle in Los Angeles, where she would work as a school principal, and he would continue in the nonprofit world—until an overabundance of backyard fruit altered the course of his life at age twenty-nine.

The couple's house in LA's West Adams neighborhood, not far from where Asante Microfarm would be years later, had fruit trees. As often happens in backyards of older homes in fertile Southern California, those trees rained down more fruit than the young family could use. So he started sharing and trading with neighbors, then gradually expanded the reach of this network through social media. A common issue, he learned, was that homeowners and tenants lacked the time or ability to harvest the fruit before it spoiled, or the birds and squirrels had taken turns eating it. The waste of good fresh food in an area plagued by food insecurity was too much for Hargins.

His solution was to launch Crop Swap LA in 2018 to harvest and care for what amounted to a far-flung urban orchard. Part of the harvest would go to the fruit tree owners, part would be given to community groups and food banks, and Crop Swap LA would sell the rest. For that, Hargins founded and ran the weekly West Adams farmers market.

Once he was in the business of turning wasted homegrown food into something productive and nourishing, it was only natural for Hargins to start thinking about what other untapped resources might be out there going to waste, which is when he started looking at grass differently. Lawns, he realized, were another kind of food waste, though instead of wasting existing food by letting it rot on the ground, this was waste of *potential* food, covering fertile soil with ornamental but useless grass. Any affection for lawns that lingered from Hargins's days of traipsing the world as a military brat vanished. His community needed that food, which meant some lawns had to go. They were food waste by omission. Hargins, who calls the lack of fresh, affordable food in this part of Los Angeles "food apartheid," sees his community microfarms as a solution and also a way of building local resilience. In addition to food supplies, Hargins says the water storage at each farm can serve as emergency water supplies during a disaster.

A month after Hargins broke ground on Asante, the lawn had been transformed into a remarkable urban farm. The boring grass monoculture was replaced by a living mural of color, textures, and variety, with rows of butter and romaine lettuces, squashes, eggplant, strawberries, bright red cherry tomatoes, beans, peppers, rainbow Swiss chard, kale, bouquets of fragrant herbs, and a pear tree filling the space. There were six hundred plants in all, sprouting from individual sacks of rich soil, compost, and nutrients, laid flat on topsoil now denuded of every blade of grass. A gently sloping farmstead in miniature, Asante is bountiful, sustainable, organic, and beautiful.

Once a week, Crop Swap staffers and volunteers come to harvest and box up the food for subscribers. The mix is different every week. Boxing day has a partylike atmosphere, and the pickup is a social occasion rather than the drudgery of a supermarket run. "Everything is so fresh, so good," one subscriber said. "I love coming by during the week and seeing what's coming next."

Once Asante was thriving, three more lawn-to-microfarm projects came next, including one for a retired, recently widowed woman who wanted to pay for the entire cost of transforming her lawn to help Crop Swap and the community she had lived in for years. This farm is smaller than Asante, but the four hundred plants packed into the beds include Desert zucchini, Cajun Delight okra, Georgia Southern collards, three kinds of kale, three kinds of lettuce and cabbage, San Marzano tomatoes, and Sugar Baby melons. There's also a pollinator garden of native plants, alive with bumblebees and butterflies, including the famous monarch with its gorgeously vibrant orange and black wings that look like stained glass brought to life. Hargins worried the homeowner might be annoyed by the regular visits from farming staff and the hubbub of the weekly harvest gatherings, but instead she assured him she was delighted. The activity and presence of what she called "my farmers" helped her cope with loneliness after losing her husband and made her feel safe, too. Hargins suggested the farm should be named for her husband, and so his second lawn reincarnation was christened the La Salle Microfarm.

Hargins has been busy: in addition to lawn-to-farm transformations, Crop Swap LA designed and built two school vegetable gardens (including one for a charter grade school with an award-winning program geared to the needs of people experiencing poverty or lack of housing), worked on a garden-science program for a STEM-focused middle school, and launched an ongoing series of "urban survival" classes on how to build an urban community farm. Dozens of other neighborhoods, schools, and communities have lined up to ask Crop Swap to transform more lawns, weedy lots, schoolyards, and dead space into flourishing farms.

Hargins's latest project is his biggest yet—a microfarm that will be a centerpiece in a major redevelopment project for South Los Angeles called Marlton Square, just a short distance from Asante. The project will include two five-story office buildings, the 19,000-square-foot Diaspora Groceries opened by actor-comedian Tiffany Haddish, and a 10,000-square-foot food incubator and commercial kitchen to nurture new food businesses and entrepreneurs, run by Black-owned food and beverage company Good Vibes Only. Crop Swap LA will operate an urban farm there, supplying fresh produce for Diaspora Groceries and the incubator kitchen. The farm will also serve as a teaching space for home gardeners and a resource for local schools.

For all its successes, the Crop Swap nonprofit is not self-sustaining—the bargain subscription prices don't even begin to cover the costs. The program is supported largely through donations and grants. A $4 million grant from the state of California gave Hargins enough seed money to staff the project, build the farms, and sustain the operations for two years, but Hargins worries about the future of the project. He has been scrambling for funds to allow continued expansion. His pleas have led to an uptick in donations, but there is no margin for error, and these days he splits his time between farming, fundraising, and worrying.

Meanwhile, his largest urban farm yet, the Degnan Microfarm (he donated his own front yard for the cause), had its first harvest in the summer of 2023, with a crop large enough to serve forty-five families a week with five-to-eight-pound bags of fruit and vegetable bounty. Degnan is Hargins's most ambitious microfarm yet—solar powered and rainwater-harvesting, with on-site composting, beehives for honey production, a medicinal herb garden, microgreens, and a mushroom farm in addition to the usual selection of veggies and fruits. With its terraced and ladder gardens, it is a traffic stopper, complex and beautiful.

"Growing your own food is so empowering," Hargins said. "Everyone has their own light bulb moment, whether it's doing something for the children or for their community. For others it's tied to their ancestry. Or maybe they just ate something nasty and decided they were sick of eating terrible food or processed food and wanted some variety or control. Or maybe they just paid a hundred dollars for getting their lawn mowed and then wonder, *Why do that when I could be picking my own tomatoes instead?*"

One day while Hargins was tending the vegetables at Asante, a neighbor stopped halfway through his morning jog to goggle.

"I can't believe the transformation," he called out. "I don't pass by here all the time, but I had to check this out."

Hargins looked up from the harvest to say thank you and chat up the visitor. Then he reached down and snapped off some leafy red stems from a large and bushy beet plant, handing one to the neighbor and one to me, and started munching one himself. The stalks looked, felt, and crunched like red celery and tasted similar, too, but sweeter. Most people just throw beet stems away, but the discarded parts of lots of vegetables are usually delicious and nutritious. That goes for the tender beet leaves, too—the whole plant is edible. Hargins took the last bite of his beet top and, bidding the jogger farewell after telling him how to subscribe, promised, "I'm just getting started."

So What's the Big Deal about Grass?

Seeing what Hargins has done to the traditional lawn, how neighbors reacted with admiration and delight rather than grumbling and complaints, puts the traditional lawn in a new light. And it is not a flattering light.

America's love of grass lawns is arguably our most perplexing yet prodigious form of habitual, unacknowledged waste. And, like a lot of other truisms of today, our lawns would have outraged the generations who came before us as absurdly and even sinfully wasteful.

Though it feels like they've been around forever, pervasive grass lawns simply were not a thing for most Americans before World War II. The lawn was an invention of the aristocracy in France and England in the 1800s, an ostentatious display of wealth and power reserved for palaces and great manors. Highclere Castle, the 345-year-old real setting for the fictional Downton Abbey, was designed with great lawns. The origin story of the American lawn is beyond shameful. Several of the United States' wealthy founders—most notably Thomas Jefferson and George Washington—were among the first to import the European grasses for their estates, in imitation of the nobility they had rebelled against, maintaining those expensive displays of manicured greenery with enslaved labor. Other Americans in those days had no interest in emulating those lawns: They cultivated food, not grass. They grew vegetables and grains. Anything else was a waste. Anything else risked starvation.

They understood what has been forgotten today: that European turf-grass doesn't belong here. It doesn't like the climate in many parts of America, which is why it is so expensive and time-consuming to keep alive and free of weeds to this day. It's an invasive species that destroys, rather than supports, native plants and creatures, anathema to natural habitats and the pollinators that we need, the bees and butterflies and birds. Evidence of lack of reverence for the lawn in the America of the past is plain even in our traditional patriotic songs. It's amber waves of grain and the fruited plain that "America the Beautiful" celebrates as iconic national symbols from 125 years ago, not our emerald lawns of grass. It was only during the homeownership boom and creation of suburbia in the 1950s that grass lawns surrounded by white picket fences shifted from a wasteful indulgence of the rich with a brutal legacy to something integral to middle-class aspirations. And we've been stuck with this expensive green boondoggle of waste ever since.

Grass lawns are and always have been impractical, costly, visually boring, and extremely needy. They are the baby that never grows up and reaches independence. Grass lawns are ornamental, a kind of landscape jewelry—except the analogy doesn't quite work. An expensive piece of jewelry, at least, has investment value that can be recovered by selling it, often at a profit. The $126 billion paid to the lawn care and landscaping industry annually—to mow, manicure, fertilize, and chemically paint grass with toxins to kill weeds and pests—offers no productive return on investment. We're paying for diamonds but getting grassy cubic zirconia instead.

Grass lawns cover about 50 million acres in the US and use 3 trillion gallons of water a year—much more water than any human food crop in the country, and the largest share of public water supplies. Got a big water bill? If you water a lawn, that's probably the reason.

In raw acreage, our grass rivals our biggest crop, corn grown for human consumption. Grass also beats the next biggest food crop, soybeans grown for human consumption (not counting the 70 percent of soy that goes to animal feed). The number of acres of grass we grow nationally could cover all the farmland in Kansas with more than 4 million acres left over—which is enough to cover every square inch of Rhode Island with lawn four times.

The lawn care and landscaping industry also consumes 59 million pounds of pesticides a year—applied to lawns with ten times the amount per acre as farmers use for crops. Children living in households where pesticides are used have increased risk of leukemia, brain cancer, and soft tissue sarcoma.

Another cost of lawns comes in the form of fossil fuels: 3 billion gallons of gasoline are used by mowers, blowers, weed whackers, and other landscaping equipment every year—the equivalent of 6 million cars on the road. Grass, like any plant, absorbs and stores some carbon, but grass is pretty bad at it. Emissions from lawn care are four times greater than any carbon removal grass manages. Grass lawns are definitely not carbon neutral.

Lawn complacency has been enduring, but cracks in the green wall are beginning to appear. People are beginning to rebel against grassy dominance, and it's not just Crop Swap LA. One movement against the lawn has been quietly building for years: a growing interest to swap out grass in favor of native plants and pollinator gardens. Of the three movements Jamiah Hargins is straddling—native-pollinator front yards, urban farming, and home vegetable gardens—the native plants seem to have the most traction nationwide. Perhaps that's because switching out a grass lawn for something that looks like a beautiful plant and flower nursery is less of a psychological leap than planting something that more closely resembles the produce aisle at the supermarket.

But its appeal means it's also spawning the most pushback. One couple in Columbia, Maryland, Janet and Jeff Crouch, found that out the hard way after they remade their large grassy front yard into a sprawling and riotous native-plant showcase. Janet, who works for the US Department of Health and Human Services, and Jeff, a clinical social worker, both commuted to Washington, DC, for work, and took joy each day in returning home to their garden sanctuary. They'd worked on it for years, advised and inspired by Janet's sister, Nancy Lawson, author of the *Humane Gardener* blog and of a native-plant garden guidebook by the same title. (It's an excellent guide to native plants for the beginner and full of interesting tangential tidbits, such as the fact that opossums, despite a reputation as pests, helpfully eat disease-causing ticks by the thousands. Who knew?) With Lawson's help and the Crouches' elbow grease, what had been a barren stretch of grass became a glorious mix of flowers, including phloxes, sunflowers, native grasses, and milkweed, which is not a "weed" any more than any other wildflowers are weeds. Milkweed also happens to be the one plant vital for the survival of the migratory monarch butterfly, and soon they were flocking to the Crouches' yard, though there was not a sign of them elsewhere in their grass-dominated, quiet cul-de-sac neighborhood. The Crouches also witnessed a flourishing of bumblebees, ladybugs, goldfinches, and hummingbirds questing for nectar. Neighbors would stop by and look on in bemused amazement. But one local was quietly simmering with resentment at the lack of conformity, at what he considered a weed-and-vermin-infested chaos—"a mess of a jungle," he later wrote. Without ever saying a word to the Crouches, he blasted their native plantings in a complaint letter to the homeowners' association.

So the Crouches were taken completely by surprise when they returned home from an outing one afternoon to find a cease-and-desist letter from the homeowners' association lawyer waiting for them, giving them ten days to tear out everything and replant grass, or face fines or more drastic action for creating an eyesore that hurt all their neighbors' property values. There had been no warning. No phone call. No informal discussion to see if some compromise could be worked out. Just threats and bluster and scare tactics. "Your yard is not the place of such a habitat," the lawyer wrote.

The couple was not cowed. Instead, they were outraged. And they fought back. They refused to change their lawn, sued, got countersued, and the whole absurd battle so angered the local state legislator that she sponsored and won passage of a new state law in 2021 forbidding homeowners' associations in Maryland from impeding native plants and pollinator gardens or ordering the planting of turfgrass. The Crouches had won. They agreed to a few small concessions to settle the lawsuits—such as keeping their plantings no closer than three feet to the angry neighbor's property line. Face-saving for the homeowners' association, but inconsequential to the Crouches. They had inspired the first law of its kind in the country.

The case drew national headlines and social media posts and had an unintended consequence for the homeowners' association—other residents in the area began following in the Crouches' footsteps, and the media attention spiked interest nationwide in pollinator gardens.

The ecological benefits of native plant gardens are numerous—bees and butterflies need all the help they can get, given that flying across a large expanse of turfgrass is for them the same as a human crossing a stretch of the Sahara. Such gardens require far less irrigation than turfgrass as well, so they avoid waste of water and money.

Home vegetable gardens and urban microfarms, which play nicely with pollinator gardens, also save water compared to what's needed for grass, and they have the added benefit of cutting food waste. In fact, growing your own produce, along with composting your food and yard waste, is one of the best things you can do to protect the environment, save money, and increase nutrition and food health. It's a win up and down the board, and it's fun and rewarding, too.

It's everything the factory farm is not, which is where the story of our enormous problem with food waste really begins.

When Food Isn't Eaten

Food waste is a global problem, but America is a standout as the world's food waste champion. Our vaunted industrial agriculture system—mechanized, pesticide-soaked, and dependent on massive amounts of fossil fuel and toxic ammonia as a fertilizer—succeeds in its primary purpose of producing incredible amounts of food. But it is also radically inefficient. Which means it wastes epic amounts of food.

Of the 241 million tons of food grown for humans in the United States in 2021, 38 percent was either unsold or uneaten, at a staggering cost of $444 billion (based on data from the food-waste nonprofit ReFED). A little of that ended up donated to anti-hunger initiatives, and some was "recycled"—meaning it was composted or used as animal feed—but the lion's share was sheer waste. Eighty million tons, 33 percent of our food, ended up in landfills.

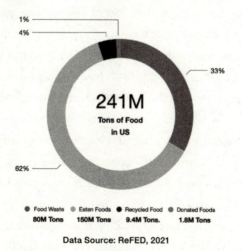

1%
4%
33%

241M
Tons of Food
in US

62%

● Food Waste   ● Eaten Foods   ● Recycled Food   ● Donated Foods
80M Tons   **150M Tons**   **9.4M Tons.**   **1.8M Tons**

Data Source: ReFED, 2021

That's the equivalent of 149 billion meals uneaten or unsold each year—about 2 percent of our gross domestic product. All the water, fossil fuel, electricity, pesticides, labor, and heat-trapping pollutants associated with growing, picking, shipping, processing, and cooking it was wasted along with that uneaten food. This makes food waste responsible for 6 percent of our heat-trapping pollutants, 22 percent of our freshwater use, and nearly a fourth of the content of our landfills (where it produces yet more of the potent climate-disrupting gas methane).

The food waste is spread out through the industrial agriculture system. Seventeen percent of food waste happens at the farm, where perfectly edible and nutritious "ugly" fruit and vegetables are considered trash, or whole harvests are left to rot in the fields because of low market prices and reduced profits should they be brought to market. Food processing and manufacturing wastes another 15 percent of the food going in, most of it from edible portions of produce cut away, such as potato peels cut from french fries or the waste from turning full-size carrots into "baby carrots." Grocery stores, restaurants, and other food services account for 20 percent of total food waste. The main driver of waste in stores is food that reaches its "sell by" date while still on the shelves, even though these foods are perfectly good and unspoiled. In the restaurant world, most of the waste is by customers who leave food on the plate.

And the biggest source of food waste? Us. Homes generate nearly half of all wasted food, whether it's left to spoil and get furry in the fridge or left on the plate. Over-purchasing is the problem, mainly because of pricing strategies that encourage consumers to buy and spend more because that's "the best deal," even though the food companies know overbuying will lead to food waste. Many people also misunderstand "expiration" and "sell by" dates, which are usually inaccurate and lead to the tossing of perfectly good food. Home cooks also are often unaware of simple cheffy tricks for storing and freezing foods to extend expiration dates, or for cooking with ingredients that seem overripe. As a result, most food waste involves perishables, with produce, prepared foods, and dairy and eggs at the top of the list.

The food we waste at home costs the average family of four $2,760 a year. Nationwide, that's $229 billion worth of moldy cheese and brown bananas and spoiled milk and leftover pizza, among other nasties in the fridge and on the counter. The heat-trapping emissions from that much wasted food every year are the same as 54 million gasoline-powered cars, and the water wasted in growing the food we trash is 11 trillion gallons—enough to fill 16 million Olympic-size swimming pools. The top three reasons people give for their food waste at home, in order: the food was spoiled; there were inedible parts (which, like the stems and leaves of Jamiah Hargins's beets, usually are edible); and nobody in the house wanted the leftovers.

Making a Personal Plan

The silver lining of food waste is that a lot of it—and all the home-based waste—can be reduced or eliminated pretty easily by improving our own choices and changing our habits.

Anne-Marie Bonneau, who blogs and writes as the Zero-Waste Chef (also the title of her 2021 book), has some great tips for avoiding food waste. Here are some of her ideas and some others not on her list, but that fit her philosophy.

1. **Make a meal plan.** "Shop" your refrigerator and pantry and make a plan to use what you have, and then make a list of the ingredients you need.

2. **Don't overbuy.** Fresh ingredients, particularly produce, need to be eaten within a week. So buy only what you need for a few days.

3. **Recipes are made to be adapted (unless you're baking).** Use recipes as guidelines, not strict rules, so your fridge and pantry don't pile up with ingredients you don't often need or use. Or just wing it with ingredients on hand in forgiving dishes like soups, stir-fries, and frittatas. Baking, however, is chemistry. Don't mess with cake and cookie recipes, because that leads to waste when they don't turn out.

4. **Serve modest portions.** Make vegetables the largest portion and animal proteins the smallest (to maximize health benefits and

minimize environmental harm). You can always go back for seconds, but once food is on the plate, if it's not eaten, it's waste.

5. **Store leftovers in glass jars and containers.** If you can't see what's in a container, it'll likely sit in the fridge, forgotten, and go bad. Also, glass is better than plastic, which can release chemicals into your food, especially if you microwave food in plastic (please don't!). Hoard your old glass pickle, jelly, spaghetti sauce, and peanut butter jars. They're great for storage, they're reusable, and they're free!

6. **Repurpose leftovers.** Fry leftover pasta and eggs in a pan, toss in chopped vegetables, and sprinkle with cheese. Leftover rice can be refashioned into fried rice (which was invented to make old, hard cooked rice and veggie leftovers into something delicious). Think like the contestants on the TV show *Chopped*!

7. **Repurpose food "waste."** Highly nutritious beet stems and leaves can be chopped and added to the mix for salads, or sautéed with a little olive oil, garlic, and salt for a vegetable side dish. Cauliflower leaves can be treated the same way in a pan or on a baking sheet laid flat and roasted until they become dark, crispy chips. Broccoli stalks should be chopped and cooked just like the florets. My favorite tactic is to freeze vegetable discards such as celery tops, carrot ends and peels (the most nutritious part), onion and garlic skins and ends, and leek tops and nubs. When you have a good amount (I use an old bread bag or cereal box liner, and when it's stuffed, it's enough), throw them in a big soup pot with water and a little salt and pepper, simmer everything for a few hours, then strain. You now have stock for any soup to use immediately or to freeze. And the stuff left in the strainer makes great compost!

8. **Make soup your go-to main dish on work nights.** A big pot of soup can provide multiple veggie-forward meals for the week. You can

prep the ingredients before bed, throw them in a slow cooker the next morning, and dinner's waiting when you get home. Don't think of a soup as a side dish—think of it as a meal in a bowl, a stew with more slurp. A salad, some cut fruit, and bread and cheese (or sliced meats) make good sides and provide some crunchy texture along with the soupy main dish. All the nutrition lands in your bowl in a zero-waste healthy meal that will taste so much better than anything you buy in a can from the supermarket. One of my favorite never-the-same soups is what I like to call summer in a bowl with seasonal veggies—to be followed by winter in a bowl as the season and the veggies change.

This is a different way to grocery shop and to think about the nature of food and food waste, an approach that avoids the habits and oversights that most often let food become garbage. And it works for working households pressed for time. Otherwise, the best-laid plans can be waylaid by convenient but expensive takeout or delivery, or cheap but unhealthy fast food, or some salty overpackaged snack we grab just to tide us over that we perceive as a mere indulgence, though we actually have made it a staple. On any given day, one out of three Americans consumes a single-serve unrecyclable plastic package of potato chips or their salty-snack equivalent—about 4 billion pounds of the stuff a year. That's a lot of chips.

The solution often proposed for bad nutrition and convenience eating usually involves some form of denial. That means a self-imposed ban or rigid diet that amounts to a just-say-no approach to the siren call of a mega burger and fries, or that bag of kettle chips we all know comes with an extra big serving of salt, fat, and carbs, even as we greedily fish out that last delicious crumb from the corner of the bag. The just-say-no approach certainly works for some people, but part of the total garbage story is recognizing that if something is both wasteful and bad for us (have you noticed how often the two go together?), yet remains enduringly popular, the best approach is not to ban or scold but to offer a better alternative. For me that's the meal soups from the home vegetable garden, supplemented by fresh goodies from the local farmers market. And the salty-crunchy reward on the side comes in the form of a few baked Parmesan crisps, which I have been pleased to learn are as satisfying as kettle chips but without the carbs and the grease. Go figure! And if you must have those snacks, just make them an occasional treat rather than an everyday staple. I don't believe in denial of things we love. But moderation is good in all things, and way easier to embrace.

This is where the home vegetable garden, the urban farm, and farmers markets come in, a new back-to-the-future synergy of food the way it used to be and can be again. Together, this trio can push factory farming and all its wasteful impacts to the sideline for a big chunk of our meals at home, with health, economic, and environmental benefits landing in the bowl instead.

The quaint backyard vegetable garden (or, as Crop Swap LA has shown, the front lawn vegetable garden) has the potential of solving a lot of waste and harm embedded in American food. And it is trending big time, with 6 million new household vegetable gardens started in recent years, driven first by the pandemic and now fueled by inflation-induced increases in food prices. Now 35 million American households are growing their own food. On average, home gardens are about 600 square feet, yielding about $600 worth of produce a year, according to surveys by the National Gardening Association. The favorite crop by far: tomatoes, followed by cucumbers, sweet peppers, beans, carrots, summer squash, onions, hot peppers, lettuce, and peas.

There hasn't been this much interest in home vegetable gardening since World War II, when Americans were urged to plant victory gardens to support the war effort (and to have food for themselves, given that supply chains for civilians were diverted to the war effort). More than 20 million households heeded the call. Which was incredible, given that there were only 36 million households in America when Pearl Harbor was attacked and the US entered the war. There were victory gardens in six out of ten American homes by the end of the war, harvesting an incredible 40 percent of the nation's produce.

What we learned then was that home vegetable gardens could both ramp up fast and compete on productivity with industrial farming. They could do this year-round and at a lower cost to wallets and the environment. Those lessons apply today. Those results can be exceeded today.

Of course, not everyone has the space or a yard for a vegetable garden. But apartment dwellers, dorm dwellers, and others without outdoor garden potential are not excluded. They are still raising bountiful potted tomatoes and other produce on their balconies or windowsills or with the help of indoor garden kits that take up little space but offer pretty impressive results. Some of those kits, available online, are surprisingly affordable and even come with watering systems and grow lights. And another million Americans with limited yard space grow produce at shared community gardens, a rapidly expanding sector of do-it-yourself food growing. We are living through a renaissance of home vegetable gardening.

Then there is the matter of taste and nutrition. Home-grown veggies win on both counts in a first-round knockout. People grow tomatoes more than anything else because the taste difference between an industrial supermarket tomato and a ripe Big Boy or cherry or heirloom tomato grown at home without chemicals and with rich home compost is beyond dramatic. Home-grown tomatoes are flavor bombs. Industrial tomatoes are *purposely* bland, bred not for flavor but firmness, flawless round skin, and durability during shipping. If you buy a winter tomato from the produce aisle, it almost certainly came from Florida's statewide tomato factory farms (it's not all oranges in the Sunshine State). Do you know what happens if you drop a tomato from your garden on the floor? You get a splatty mess, because a tomato ripe with juice and nutrition does that. It's supposed to do that. The very things that make it juicy, flavorful, and nutritious make it fragile. But drop a "perfect" Florida winter tomato you just brought home, and it will roll away, unharmed.

And what has sixty years of industrial farming of tomatoes done besides giving tomatoes a perfect complexion and the delicacy of a golf ball? Well, the US Department of Agriculture has test data for that. Today's industrial tomatoes have 30 percent less vitamin C, 30 percent less thiamine, 19 percent less niacin, and 62 percent less calcium than a vintage

1960 tomato. The only thing today's industrial tomato has more of is the one thing Americans don't need: fourteen times more salt.

Industrial farming has sapped the nutrition out of most of our staple foods. Wheat and barley today have up to 50 percent less protein concentration as they did before World War II. Protein, oil, and three vital amino acids have all declined in corn since then, too. Calcium in broccoli produced on industrial farms is a third of what it was in 1950—in a vegetable renowned for being rich in that vital mineral that builds strong bones and teeth. Your vegetable garden produce does not suffer from this nutrition-robbing malady. Home-grown is healthy and planet friendly.

But a mountain of scientific evidence shows our modern industrial fruits, vegetables, and grains carry less protein, calcium, phosphorus, iron, riboflavin, and vitamin C than the same types of food our grandparents grew up eating. What's more, the same effect has diminished the nutritional value found in feed for cows, pigs, goats, and lambs—which means our meats are less nutritious, too. And the culprit for all this is the key industrial farming fertilizer of the modern world, ammonia—a powerful source of nitrogen fertilizer that promotes plant growth but that also is a caustic, corrosive chemical poisonous to touch, breathe, or swallow. Ammonia is the secret to modern farming's success, but it's the reason why industrial farming's vaunted productivity is ruinously wasteful, wreaking havoc with our diets, our soil health, the environment, and climate.

Big Pharm

This conundrum began in 1909 when German chemist Fritz Haber discovered the holy grail chemists had sought for half a century: how to make artificial fertilizer out of the most plentiful gas in the atmosphere, nitrogen, a substance that plants need to grow but can't use in its airborne form. Haber figured out how to combine it with hydrogen under extreme heat and pressure to make ammonia, which could be made into a super fertilizer in place of the age-old, less intensive natural method of using animal manure. His tabletop experiment was scaled to commercial proportions four years later by another German chemist, Carl Bosch. The Haber-Bosch process, still in use today, was hailed as a breakthrough and a solution to world hunger. Spraying farmland with ammonia could radically increase the amount of nitrogen in soil, and it more than doubled the amount of food an acre of farmland produced with traditional methods. Haber received a Nobel Prize in Chemistry in 1918 for discovering the ammonia synthesis process, and Bosch received a Nobel Prize in Chemistry thirteen years later for developing high-pressure manufacturing methods critical to mass production of ammonia fertilizer. Together their work drove the modernization of agriculture, magnifying the productivity gains of an era later dubbed the "green revolution."

That title became an ironic one, however, as the downsides of this wondrous breakthrough became clear many years later. Ammonia became the most manufactured chemical in the world, but the Haber-Bosch process is so energy-intensive and so reliant on enormous amounts of fossil fuel that it consumes 1 percent of the entire world's energy production. As a result, the process of manufacturing this one chemical is also responsible for 1 percent of global warming.

"Ammonia as it's produced today for fertilizers is effectively a fossil fuel product," observed Douglas MacFarlane, a research chemist at Australia's Monash University. "Most of our food comes from fertilizers. Therefore, our food is effectively a fossil fuel product. And that's not sustainable."

That's putting it mildly. The other big problem is that ammonia is not just any poison. It is, paradoxically for a fertilizer, so poisonous that it is incompatible with life in its raw form. Farmers are advised to wear chemical-proof goggles at the least, though respirators are suggested, too, when applying the caustic substance, which can burn, blind, and destroy lungs and any other soft tissue from accidental exposure. It has a similar effect on soil, which is why it goes on before any planting, allowing it to be absorbed and have time to evaporate, leaving usable nitrogen behind. On contact, it can kill everything, even the microorganisms and fungal networks beneath the surface—what is, in normal naturally fertile land, the hidden magic that distinguishes dirt from soil. The underground web of life is symbiotic with plant root systems and helps growing crops absorb all the minerals, vitamins, and other nutrients that make veggies and fruit healthy and delicious. Ammonia damages that, turning soil back into dirt, and so our industrial crops are by design less nutritious. Yes, we make more food by weight and volume than before this poisonous revolution. It's just not as good as the food produced with natural methods. The soil of modern agriculture is often dead. We grow our food on nature's corpse.

The other hallmark of industrial agriculture, chemical pesticides, completes the process. The combination of chemicals is not contained to the field where it is sprayed, but rather is spread by rain, runoff, and wind. It pollutes rivers, lakes, and groundwater supplies used for drinking water and irrigation. It harms ecosystems worldwide and has created marine dead zones where fisheries once thrived in the Gulf of Mexico and other bodies of water around the world. Some of the ammonia is converted in the fields to nitrogen oxides, the same toxic air pollutants that stoves emit and one of the most potent heat-trapping pollutants, three hundred times worse than carbon dioxide. This is one of the reasons agriculture is the biggest climate disrupter globally, though in the United States it is responsible for just 10 percent of our climate impact. America is one of the minority of countries in the world where giant transportation and energy sectors dwarf farming's heat-trapping pollutants. Globally, our farming and food supply chain is responsible for at least 26 percent of human-generated heat-trapping emissions, with some estimates putting that figure at 35 percent.

Animal agriculture is the biggest culprit when it comes to the environmental devastation caused by factory farming, driving nearly 60 percent of food's climate impact, as well as being a major force in worldwide deforestation. So meat and dairy are a climate-killing double whammy: they spew worldwide emissions worse than all the cars in the world, while cutting down nature's primary means of absorbing those same climate-altering gases.

We grow more food for our livestock than for ourselves. It's the most

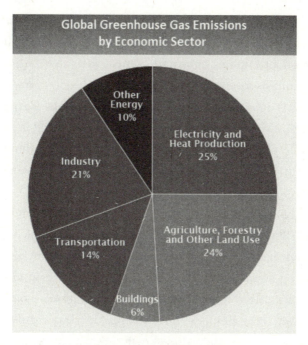

Source: U.S. Environmental Protection Agency

heavily subsidized part of the US food economy. One out of every three dollars in federal farm subsidies go to feed crops for dairy, meat, and eggs—most of which support giant and extremely profitable factory farming corporations, not struggling family farms. As for fruits and vegetables, which federal dietary guidelines, food and nutrition science, and simple common sense tell us are the single most important part of the human diet? Fruits and vegetables get only 4 percent of federal agriculture subsidies.

That imbalance isn't the only thing wrong with this picture. We are spending the most money on the least nutrition. Seventy-five percent of US cropland feeds animals, not people. It's a terrible trade-off because animal agriculture uses 83 percent of the world's farmland to produce only 37 percent of our protein and 18 percent of our calories—even as it churns out 60 percent of food's climate impact. It's basically the food transfer version of gas stoves' heat transfer inefficiency. Growing food for animals, then eating the animals, is far less efficient—and less nutritious—than growing plant nutrition for ourselves to eat directly. The transfer lessens nutrition while increasing environmental damage. We should be subsidizing plant food for humans more and livestock less.

But Americans love their meat. It's a habit. It's on every major street in the country: subsidized, carbon-heavy fast-food meat. And of all the animal proteins we consume, the very worst for the world is beef, America's favorite food and worst eating habit. It is absolutely the worst thing you can eat in large quantities for the health of planet, people, and the economy.

By weight, beef generates three hundred times as much climate-disrupting pollutants as asparagus, fifty-seven times as much as potatoes, and thirteen times as much as the equivalent weight of chicken—the least climate-killing animal protein in our diet. Every half-pound hamburger you eat carries emissions equivalent to driving a three-thousand-pound gasoline-powered car 10 miles. The average American's annual beef consumption carries about the same climate impact as six months of that same American's driving.

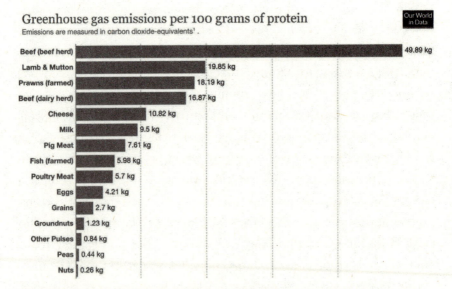

Greenhouse gas emissions per 100 grams of protein
Emissions are measured in carbon dioxide-equivalents[1].

Beef (beef herd)	49.89 kg
Lamb & Mutton	19.85 kg
Prawns (farmed)	18.19 kg
Beef (dairy herd)	16.87 kg
Cheese	10.82 kg
Milk	9.5 kg
Pig Meat	7.61 kg
Fish (farmed)	5.98 kg
Poultry Meat	5.7 kg
Eggs	4.21 kg
Grains	2.7 kg
Groundnuts	1.23 kg
Other Pulses	0.84 kg
Peas	0.44 kg
Nuts	0.26 kg

Our World in Data

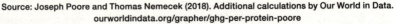

Source: Joseph Poore and Thomas Nemecek (2018). Additional calculations by Our World in Data.
ourworldindata.org/grapher/ghg-per-protein-poore

This is the food paradox: we have national policies aimed at lowering our collective climate emissions and promoting healthy diets, yet we have massive farm subsidies that prop up the most carbon-intensive, environmentally damaging, and least healthy foods—actually giving more money to subsidize animal feed than human food. At the same time, we completely shortchange the healthiest, most nutritional, least environmentally harmful foods available, fruits and vegetables. Of the $424 billion in US farm subsidies, about half went to just three crops: corn, wheat, and soybeans. Most of the corn and soy became animal food. The next main use of corn in the US is to make corn syrup, the cheap and unhealthful sweetener in processed and junk foods, and to make ethanol, a gasoline additive that delivers little or no environmental benefits. Incredibly, even as corn gets a whopping 27 percent of all federal farm subsidies, only about 1 percent of the subsidized corn is actually eaten by humans in its nonprocessed, healthy natural form.

This is a form of systemic food waste very different from not cleaning your plate or letting food get furry in the fridge. Our food system is defined by wasteful priorities and politics that allow subsidies to distort the food market into prioritizing the worst kinds of food, which, if not so heavily subsidized, would not be financially viable. This is the very definition of total garbage.

And it is here that we have the ability as consumers and parents and communities to use our power of choice to overcome these perverse incentives. What we choose to eat has profound effects on the climate. Shifting toward an increasingly plant-based diet carries huge benefits in the war on waste and the crises it drives. A vegan diet can cut food-related carbon emissions by nearly 60 percent. And even a healthy Mediterranean diet with moderate amounts of meat can knock off 20 percent of your food's climate impact. That's huge. That's as big an impact as electrifying your home if you're currently burning fossil fuels. That's worth considering—especially because Mediterranean food is absolutely delicious.

Again, banning things we love is a lousy way to change the world, mostly because it doesn't work. For me, making meat a treat, not a staple, works well, a decision I made for health reasons originally, but that now makes me feel good for environmental and humane reasons, too. I simply feel better eating a mostly plant-based diet. My gut is healthier, my energy is higher—it's a win-win for me. But now and then a good steak is a great treat; I'm partial to teppanyaki—who doesn't like dinner and a show?

Zero-Waste Chef Anne-Marie Bonneau takes a similar approach to her diet: plant-forward, with her animal protein mainstay being eggs—the lowest climate footprint of all animal foods. She says she's always finding new ways to incorporate eggs in meals. "I love eggs. I eat quite a few. I made sourdough pancakes yesterday, and I happened to have duck eggs. Soooo good."

So, one way to approach this is to focus on the easiest of the Eight Rs: reduce. If you eat meat four times a week, try making it two or three instead, and see how that works for you for starters. If you mostly eat beef, maybe shift the balance to more chicken or pork. If you're a cheese fan, some of the nut-based cheeses, particularly the spreads, are delicious, and

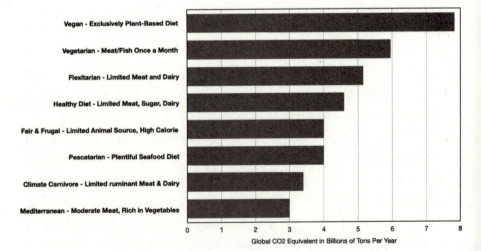

Annual Global Greenhouse Gas Emissions Avoided by Diet Change

(Current world food production emits 13.7 billion tons of greenhouse gases, 1/4 of global heat-trapping pollutants)

Data from United Nations' Intergovernmental Panel on Climate Change
Climate Change and Land, August 2019

plant-based milks are much better environmentally, too. I've been using an almond-macadamia-cashew mix that is great in cereal, and I found an oat milk creamer that goes pretty well in coffee. Give a few a try, see if any work for you.

The only reason that our distorted, heavily subsidized, and wasteful industrial agriculture system can exist affordably—and the reason its advocates can claim it is the most efficient and only possible way to drive the $1.1 trillion US food system—is because it cheats. In addition to the billions in overt subsidies, the food industry has a larger invisible subsidy: it's excused from responsibility for the cost of its damage to water and air quality; for its devastation to habitats, fisheries, and biodiversity; for the health-care costs that unhealthful food imposes; and for the price tag of climate disruption and extreme weather related to its heat-trapping pollutants. A study by the Rockefeller Foundation put the cost of those "externalities" at *$2 trillion*—nearly twice the market value of the industry itself. It's the economics of plastic pollution all over again. Industrial agriculture corporations wreak the damage, but the public bears the cost. This is yet another form of food waste, the massive shadow waste that doesn't show up at the landfill, but is spread throughout our living world.

"Americans pay that cost even if consumers don't see it at check-out," the report's authors concluded. "And if we don't change our food system, future generations will, too."

This is why our local farmers markets, urban farms, and backyard vegetable microfarms are so powerful, with their low-impact, low-cost, and high-quality food. They are, with an accounting that takes the entire balance sheet of benefits and costs, *more* efficient than industrial farming. Supporting the farmers markets and urban farms as a customer, while building your own version of a farm at home, even if it's just a few pots of tomatoes and basil for starters, is one of the most effective waste-cutting and world-healing things you can do—while reaping the benefits of great affordable food.

If there is a single thing you choose to do to create a world that is not total garbage, this is a great choice. And a tasty one.

Taking Big Farms Back to Basics

The good news is that other forces are trying to counter the notion that large-scale farming has to kill soil before it can be used. It's called regenerative agriculture, and it is growing as fast as home vegetable gardening. Regenerative farms are embracing the opposite of industrial farming methods—making soil health their guiding star—and they are making the case that they are the ammonia-shunning future of food. If home vegetable gardens and urban farms are the rooftop solar installations of the food world, then regenerative agriculture farms are the renewable-energy power plants taking on the fossil fuel incumbents.

Regenerative agriculture has proved its value in side-by-side tests with fields farmed with conventional methods, using advanced soil science combined with commonsense traditional farming methods relied on long before Haber-Bosch came along. Those commonsense things seem almost ridiculously obvious and include the less-than-radical notion that poisoning the soil you farm is not a good long-term strategy. Regenerative farming does reject, or at least avoids as much as possible, one age-old farming technique: tilling the soil. That turning over and breaking of the soil depletes minerals and disturbs the organic life beneath. It also releases all the carbon dioxide that plants and soil microorganisms so helpfully sequester there for us. Regenerative farmers leave the soil alone and protect it with cover crops such as clover, ryegrass, or hemp to stop weeds and erosion between plantings of money crops. No-till farming also turns farms from carbon emitters into natural carbon-storage facilities; the heat-trapping pollutants stay in the ground. It also retains moisture better and makes crops more resilient in the face of climate disruption and high temperatures.

Farm animals are also welcome in regenerative fields, adding to the natural fertilization of the soil. Industrial farming fights nature. The goal of regenerative agriculture is to restore soil health by farming in harmony with nature. Pollinators, birds, and bees thrive on regenerative farms, which avoid using toxins or synthetic chemicals whenever possible (and when they are also organic operations, that avoidance is absolute). "We let nature do its thing," says Steve McElroy, who turned his 800-acre former crop farm in Michigan into a regenerative sheep and cattle ranch. He says that the farm is thriving and that the beef is healthier and more nutritious than the meat from factory farms.

The ability of regenerative farming to resurrect the nutrition of the past has been confirmed in the lab. Analysis of cabbages grown regeneratively and cabbages grown with ordinary organic methods in different fields at Singing Frogs Farm in Sebastopol, California, showed significantly more nutrition for the regenerative produce. The regenerative cabbages had 46 percent more vitamin K, 31 percent more vitamin E, 33 percent more vitamin B_1, 60 percent more vitamin B_3, and 23 percent more vitamin B_5, as well as more calcium, potassium, carotenoids, and phytosterols.

Field tests by the Rodale Institute in Pennsylvania have shown that regenerative methods, after a transitional couple of growing seasons to restore the soil health of chemically farmed fields, often have similar yields—and higher profits—than industrial farming. For some crops, the yields might be lower, but because there are none of the expenses of pesticides and ammonia fertilizer, the profits per acre are still higher than those of industrial farms.

Rachel Schattman, an assistant professor of sustainable agriculture at the University of Maine and a farmer, says the research is compelling, and that regenerative farming is going to be increasingly crucial to protect our food system in this new age of climate disruption and high temperatures. It's also a means of turning an industry from being a major driver of climate disruption into one that is more than carbon neutral and instead actually absorbs and stores harmful heat-trapping pollutants in the soil.

Schattman is working with major brands, such as General Mills, that are investing in regenerative agriculture and helping farmers make the transition away from conventional toxic industrial agriculture.

"We know enough right now to support farmers as they adapt to a changing climate, build resilience into their farm operations today, and anchor thriving US agricultural industries," she testified before a US Congress hearing on the future of farming. "We can do this through unwavering support for sustainable, regenerative agriculture."

Pocono Organics in Pennsylvania is a good example. The farm started as the passion project of owner Ashley Walsh. She was diagnosed with gastroparesis, a paralyzed stomach, in her twenties, which made digesting food difficult and painful. Feeding tubes and lifelong medication were the traditional courses of treatment for her condition, but Walsh found the results and the limitations mixed at best. She had heard that others had been able to manage the illness with an organic, chemical-free diet, and she gave it a shot. Eventually she was able to construct a diet her body could handle without feeding tubes or heavy-duty meds.

"Life-changing" is how she described it, but frustrating as well, as sourcing the good organic foods in Pennsylvania's mountain country where she lives wasn't easy. So she asked her family for a few unused acres near the family business, the Pocono Raceway.

What started as 50 acres in the town of Long Pond in 2018 is now 381 acres and North America's largest regenerative organic-certified vegetable farm, as well as the nation's first regenerative hemp farm. A four-story-tall, 40,000-square-foot solar-powered greenhouse allows year-round growing, and Rodale has opened a regenerative research center there to study the quality and productivity of a farm attempting regenerative agriculture techniques on this scale.

"The environment of a regenerative farm is unlike any other," Walsh says. "The fields are beautiful. We think this is the future of farming."

9

The Car of the Future Isn't What You Think

The mayor zips up to city hall in silence and parks in one of the premium spots in front—spaces reserved not for local leaders but for any driver of a special vehicle like hers. Her cheeks are ruddy and her blond hair ruffled from the ride over, which included a long, scenic stretch by the town lake, with a short pause to bird-watch.

"Not bad, right?" she asks. "Another day's commute, another day without traffic. That's why I wouldn't want to live anywhere else."

If there's a touch of schadenfreude behind Mayor Kim Learnard's exuberance, it's more than justified. After all, hers is one of the few cities in the world with a personal mobility solution that seems to check all the boxes. The system here has stripped away waste and inefficiency and has an impeccable safety record, absolutely no gridlock, no smog, no heat-trapping pollutants, and no tolls. It's more convenient than public transit and way greener than owning an electric car, at a fraction of the price. And how do her constituents feel about transportation here? "Wouldn't live anywhere else," a server at the local pizza place says. "I feel bad for people who don't have this."

The mayor lifts up the rear seat to show off the bank of six inexpensive old-school lead-acid car batteries that power her ride, the local vehicle of choice, her trusty NEV—neighborhood electric vehicle—otherwise known as a golf cart.

Yes, that's right, the true future of transportation is the lowly electric golf cart. Locals have been happily proving this to be true in Learnard's city for decades, long enough to make it normal and unremarkable to them, though it's astounding to pretty much everyone else. Give up a car? For a golf cart? Seriously?

Just think about it, Mayor Learnard suggests to the doubtful and the incredulous. How far are most of your car trips? With how many people in the car? How much cargo? Then think about what kinds of vehicles can handle most of your trips. Not all of them. Just most of them.

It so happens the government has a stat for that: half of our vehicle trips are 3 miles or less. Even a golf cart, which tops out somewhere between 25 and 35 miles per hour, can cover that distance in less time than it takes to microwave the average frozen pizza. And the vast majority of our trips are only 10 miles or less—right in the golf cart sweet spot for time, speed, and distance.

To Learnard, that means a street-legal electric golf cart is the perfect vehicle for most people most of the time: lean, clean, and green transportation that can be bought used for $4,000 or $5,000, or new for about twice as much, or more with lots of extras. Learnard's town of about thirty-five thousand people evolved to accommodate this alternative form of transport with a secondary local road system built of twelve-foot-wide asphalt mini-roads winding around neighborhoods, shopping and business zones, local schools, parks, and wooded areas. No regular cars allowed.

This zero-waste transportation system of the future was started way back in the 1970s—part of the infrastructure has always been there, though it has multiplied over time to over 100 miles of local golf cart roadways today, linking every part of the city. But it's not an either-or thing. The primary streets and roads are still present, though occasional well-marked crossings pop up on local highways like deer-warning signs, only these yellow caution signs feature a cute little golf cart silhouette trundling by. Most residents still have regular cars and pickup trucks for longer trips or hauling lots of people and stuff. They just don't need them most of the time. And families that might have had two or three full-size cars in other cities do just fine with one. Some households have ditched cars entirely.

"People here spend a lot less on transportation," Mayor Learnard says. "Also the views on our pathways are beautiful, with the lakes and the woods. Cars are solitary. These golf carts are for something different, for taking the kids out for pizza and ice cream on a Saturday night. For driving to school. And they are just fun to drive. They are social."

Transportation planners from around the world have come to visit, hoping to apply the lessons learned back home. But the near-universal reaction of just about everyone else to this kind of talk from Learnard is incredulity: *Sure, you could handle a few things with a golf cart you might otherwise do with your car, but be serious. In the end, they are just terrible substitutes for a* real *car.* That's the general truism people bring with them, Learnard says. But then you tool around a bit in her low-tech, low-speed, low-stress golf cart, and before you quite realize it, the grip of old habits and creaky assumptions loosens. And suddenly you get it, and the truism flips on its head. Given the way we actually use them day-to-day, the real problem with our cars is that they are terrible substitutes for golf carts.

"I'm there," Learnard says with a laugh. Her whole town's been there for years.

So where is this city of modest and green transportation? Some epicenter of environmentalism and alternative transportation—perhaps in Denmark or Norway or the Netherlands or Singapore? Or maybe a new theme-park city of the future like Disney's Epcot? A retirement community? Nope, none of the above.

Kim Learnard is mayor of a town 30 miles south of the sprawl and traffic of Atlanta, a place called Peachtree City, Georgia, where the motto is "Plan to Stay," and the locals are waiting for the rest of the world to catch up.

Looking at cars in a new way is hard. I get it because, though I don't like what cars have become, I do love to drive. I always have. I've loved it since

I turned fifteen and a half and got in line at the DMV for my learner's permit that same day. I loved my wreck of a first car, a battered old Chevy Nova I bought used in an auction from the city of Philadelphia for $240; it successfully got me through my last two years of college and to my first newspaper job halfway across the country. But I was really hooked the first time I successfully completed a turn without the beginner's death grip on the steering wheel, instead loosening my hold at just the right moment so it could spin back around as the car straightened out all on its own. This seemed miraculous to sixteen-year-old me, and the memory is still sharp and deep—the feel of the old family car's steering wheel whispering against my palms and fingers as it moved by itself, the first of uncountable times. There's an actual term for this. It's called self-centering, which sounds very zen, though engineers will tell you it's just basic physics—torque and centrifugal force doing their thing. I knew a higher truth in that moment, though: driving was magic. "Did you see that?" I asked, turning to grin at my dad. "Eyes in front," he barked, but he was smiling, too.

Lots of memories and little moments are wrapped up in our cars—loved ones we've lost, rites of passage, first dates, drive-throughs, drive-ins, and drives down the shore. So much engineering genius and design brilliance, generations of it, have gone into making our cars what they are today, to make those memories possible. But we've made them too central to our lives. Cars aren't freedom. They aren't irreplaceable. They are way too expensive. They are tools—just like bikes are tools. And trolleys. And a lot of the time, cars really are not the best option for getting around. This was understood when cars were new and public transit was far more robust in many cities than it is now. We've forgotten that it was never supposed to be all about cars.

Before I loved driving, I loved my hometown Philadelphia's extensive public transit system: buses, subways, the El, and my favorite, the trolleys. I'd go out of my way to ride one instead of the bus. And at night in bed, if the wind was right, I could hear the clack of the metal wheels on the tracks and the shushing rainwater sound the trolley pole made as it slid down the overhead power line. I knew every inch of that transit system from when I was a little kid. For some trips, transit was the better, quicker, and certainly cheaper option to get through the city, particularly during rush hour. Now, that's freedom, or so my child self thought.

I also loved my bike—both before I learned to drive and after. I went everywhere on it, from the bustle of Center City (what Philly residents call the downtown area) to outlying parks and rural roads, even across the bridge to New Jersey, where I had a summer job at a trucking company. As a teenager I joined the American Youth Hostels and took my first vacations without parents—by bike. It was a cheap way to travel, though I still had to save up all year to afford it. One summer I biked across Cape Cod and Martha's Vineyard with a group of kids. The next summer it was a long camping and biking trip from Boston to Deer Isle, Maine. I've never been more fit in my life. Those biking memories are as indelible as any car trips, too: meeting people from all over, visiting new places, and actually *seeing* those places because the pace of a bike trip allows that—demands it, really. You miss so much behind the windshield of a car at 60 miles an hour. So before I loved driving, I loved biking and transit, and still do.

But this is America, and cars are inescapable. And as a longtime resident of Southern California, where public transit is more of a struggle than it ought to be, I haven't found a way to get by without using a car at times. I also get why many people find cars alluring. They are, without question, genuine technological marvels, thirty thousand parts all coming together (three times the number Henry Ford used in 1908 to build his revolutionary Model T), and that's just the mechanics. There's also the army of microprocessors that have made our cars into rolling computers in the last decade or so, 1,400 semiconductor chips at minimum, up to 3,000 in some cars, and 100 million lines of code to tell them what to do. That's remarkable considering a commercial airliner might have about 14 million lines of code, and your average personal computer a few dozen semiconductor chips. So our cars are genius.

What's Really under the Hood

If only the story ended there. But it does not: the auto industry has devoted all that brilliance, talent, and capital not into making something original and great. Instead it's all about propping up an obsolete status quo, to hang on to something that no longer does what we need it to do. As a result, what started out as revolutionary, the machine that built our modern world—or more precisely, the machine we built our modern world around—is now something awful. Our cars are terrible, and there is no shortage of reasons why. The real question is why we put up with it.

If any other device in our daily lives wasted 80 percent of its fuel—priced at $5.59 per gallon in Southern California as I write this—we would look for any excuse to dump the thing. Except, when it comes to our cars, we look for any excuse *not* to try something else.

Any other product with emissions so toxic they would kill us in minutes if they accumulated in our homes would be perceived not as an object of desire but as a curse and an outrage. And those emissions don't have to be blown into our houses at lethal levels to get us in the end. A Massachusetts Institute of Technology study has confirmed our cars are killing us slowly, too. Each year, long-term breathing of smoggy air caused by cars leads to fifty-three thousand premature deaths, the victims robbed of, on average, ten years of life they would have enjoyed in the absence of tailpipe emissions. We've known this for decades. It is undeniably terrible. Yet our cars abide.

Then there's the economics of cars. For a product with such problems, cars are costly things—next to housing, our *most* costly thing. Add up the expense of fuel, insurance, maintenance, financing, and depreciation—unlike our houses, our new vehicles begin losing value from day one—the average new car today costs around $10,728 a year to own and use, according to the American Automobile Association's annual calculation of the cost of car ownership. If you drive a half-ton pickup truck—the most popular personal vehicle in America—15,000 miles a year, the cost goes up to $12,932 a year. We spend all that for a car that sits parked 92 to 96 percent of the time—basically twenty-two to twenty-three hours a day of idleness.

It is literally a terrible value proposition from a user point of view, and those few percentage points spent on the road carry gargantuan costs.

Which brings us to the most visceral, immediate damage wrought by our trusty cars and pickups. If any other product we owned killed more than 46,270 of us a year, while injuring another 3.4 million seriously enough to land them in an emergency room, we certainly wouldn't shrug it off as the price of doing business. This would be headline news, a call to arms; there would be congressional hearings demanding action if it were airliners or microwave ovens or hair dryers. We've grounded entire fleets of aircraft, civilian and military, for fewer deaths than caused by one day of driving in America. Only with cars are we so conditioned and numbed to the unintended carnage by habit and the belief that this waste of life is just the way it is and always will be. Some will even tell you, "But wait, we collectively drive 3.2 *trillion* miles a year—the staggering equivalent of five thousand round trips to the moon—so those deaths are proportionately quite low." Yes, defenders of car orthodoxy make that callous argument, but it's the wrong measure. Look at deaths as a function of population, not miles, for a true picture of how dangerous something is. According to the Insurance Institute for Highway Safety, the death rate for driving in America is currently 13 deaths per 100,000 people—which makes driving about as deadly as working construction, operating mining machinery, or being a cop. Cars in their current form pose this question: Should driving your kids to school impose the same risks of injury and death faced by police officers sworn to protect the public and who go into harm's way to do so? Police officers are paid for their professional duty and risk. Drivers pay to face the same level of risk on the road.

So what does that look like on any given day? Fatal car crashes in the United States in 2022 averaged 127 traffic deaths a day, which works out to one fatal crash every eleven minutes. In the time it takes you to unload your dishwasher, the odds are that someone will have died in a car crash somewhere in America. As for the injured, that number averages out to one trip to the emergency room every *ten seconds*. Around the clock, every day.

The costs of all that automotive waste—lives, property, police, para-medics, road closures, traffic jams, and medical care—is estimated by the National Safety Council to be a half trillion dollars a year. Those 3.2 trillion miles, it turns out, are very, very expensive. That's enough money to give every household in America with an income of $50,000 or less a really nice free NEV or an $11,000 credit to buy a full-size EV. It is enough to cover the entire cost of the thirteen-year-long Apollo moon program—twice. Three years of road-carnage costs could finance currently planned multiple manned missions to Mars and the establishment of a permanent settlement on the Red Planet. Or it could buy every single US household an NEV or give everyone that $11,000 EV credit. Switching most of our driving tasks to NEVs would radically reduce traffic deaths of both car occupants and pedestrians—because of the lower speeds and lighter vehicles, there are fewer wrecks and less damage. It's not rocket science.

Then there's that sobering Lawrence Livermore National Lab flow-chart that tells us transportation use of petroleum products in our fossil fuel vehicles is our most wasteful consumer of energy. The value of the petroleum wasted yearly by our fossil fuel transportation system, assuming a value of oil at $80 a barrel, is $292.4 billion. Our cars use nineteenth-century technology for their propulsion system—exploding gasoline inside metal cylinders that forces pistons up and down, which turn the crankshaft, which is attached to a gearbox, which in turn transfers that motion to the wheels to make them spin. That's how cars work now, and it's how they've worked from the beginning. It's the primitive heart of the car and the heart of its waste and inefficiency. Yet we have let what was undeniably revolutionary in 1900 persist far past its expiration date. If we stuck with everything that long, we'd still communicate long distance by telegraph, use heroin and toxic mercury as medicines, conduct surgery without masks, gloves, or washing our hands, and stuff our buildings full of cancer-causing, lung-destroying asbestos because it's a cheap and readily available building material. We don't do any of those things anymore for one simple reason: they are terrible. Yet somehow this archaic car technology that does all that bad stuff remains untouchable.

That's why our cities spend enormous amounts of time, energy, and money trying to make traffic better and thereby encourage *even more* driving, which of course makes traffic worse, along with the endless loop of smog and frayed patience that we never seem to learn from. If we stopped engineering our communities on behalf of cars and did something truly radical—employing traffic design to *discourage* cars in the parts of cities and towns where people could be walking or biking or driving golf carts or taking transit instead—that would be the dawn of a new age of enlightenment. Flipping that switch would save energy, emissions, and lives while making cities more livable and joyful by reinvigorating city street scenes. Seattle managed to do this by creating special bus lines for "RapidRide" transit to cut through traffic congestion while cars crawl along in rush hour madness. Getting to downtown Seattle by transit is faster, cheaper, and more pleasant than driving, with none of the parking hassles—which was a purposeful choice by city planners. It worked. More people take public transit to work in downtown Seattle than drive alone in their cars. That shouldn't be the exception. It should be the rule. Yet few cities have the guts to pull that particular trigger. But think about what the Seattle experience tells us about cars: the less they are used, the happier we get, and the easier it becomes to get where we want to be.

All this should be more than enough to wake us up to the reality that our cars, genius engineering aside, are total garbage, but there's also yet one more thing. These energy wasters, along with other fossil fuel vehicles, are a major source of heat-trapping pollutants globally, and the single largest source of climate disruption made in the USA. As I was working on this chapter, the most widespread and deadly heat waves in history gripped the country, with 60 percent of the nation's population under an extreme heat advisory and the city of Phoenix on its thirtieth consecutive day of temperatures hitting 110 degrees. Children who accidentally fell on the surface of the streets there were so severely burned they needed skin grafts. This is not normal. Or, rather, it wasn't normal until now. Our cars are a major driver of this.

The economic costs of vehicle air pollution in America, both toxic and heat-trapping, are estimated to be $180 billion a year. Add that to the half trillion in costs from car crashes and the $292.4 billion in wasted energy, and the total resources trashed in a single year of American driving rounds up to an almost inconceivable $1 trillion. How to make sense of that number? Well, our transportation waste is bigger than the national revenue of every country in the world except for five—the United States, Japan, China, Germany, and France. Everyone else in the world gets by on less money than our vehicles waste.

Strip away the nostalgia, the habit, the high-tech shell the genius engineers have slapped on the surface of our gas-guzzling internal combustion behemoths, and just look at the numbers: our cars are worse than terrible—they are stupid. And we have been stupid about our cars for a very long time. Myself included.

This is not just the fault of the fossil fuel engine. That's *a* problem—the easiest part to fix—but it's not *the* problem. The core problem is the entire design of the car, which beneath the glitzy new tech is based on fundamental principles created for the America and the world of 150 years ago. Today's cars are still stuck in a design ethos perfected for horse-drawn carriages: a rectangular box on four wheels wide enough to be pulled by two horses side by side, but not too wide that you couldn't efficiently use it with just one horse. The first cars were literally horseless carriages, with motors replacing horsepower, and steering controls instead of reins. And we've pretty much stuck with that design and those proportions ever since. Actually, the design was quite flexible during the horse age. You could add a second engine—another horse—for more power, or a team of horses for monster truckloads. Or if you just needed to transport one person and only a modicum of cargo—picking up a sack of groceries, let's say—you could unhitch the wagon and just ride the horse, the biological equivalent of a car that morphs into a motorcycle. Now, that would be cool, because that would actually be a design revolution. But we're not able to morph matter yet, so the big box on wheels is what we have.

And that design worked brilliantly for the first mass-produced, mass-affordable car, the Ford Model T, in part because it was far lighter, weighing in at only 1,200 pounds, and more nimble than today's vehicles. Today's average car is obese at 4,329 pounds, quite a bit over two tons, with SUVs flirting with and often reaching or exceeding three tons. Even the average compact car in America weighs more than two Model Ts, which were designed for a mostly rural America with poor dirt roads and cities with dirt streets. They were able to traverse streams and rattle across fields and pastures. And the Model T *could* morph: there were kits for turning one into a pickup truck, a tractor, or a plow, and even kits for taking off a rear wheel and putting on a pulley hub to power a band saw, a pump, an electric generator, a conveyor belt to lift corn into silos, or a threshing machine to harvest grain. That versatile design met real needs of the day, which is why the record it set for numbers of cars sold beat every other model of every brand of car for more than sixty years, long after it went out of production. It was not dethroned as America's best-selling car until 1972, by another lightweight, dirt cheap car, the original and now iconic VW Beetle.

Today's cars are built big, they don't convert to anything, most are unlikely to do well crossing rivers without a bridge or easily powering a corn-silo conveyor belt, but they are far more luxurious and well appointed, and their safety features are vastly improved compared with those in older cars. The old wagon-derived design parameter persists, but with all the flexibility and economy removed, which means every time you drive a car, you are tooling around with the power and space to carry five people, plus their suitcases or whatever other cargo you want to carry, and capable of going 300 miles on a full tank of gas. Our cars are designed to go long distances at high speed—on average, about 100 miles per hour, way higher than is legal, which explains why the most common cause cited in all those fatal crashes is speeding. (The Model T maxed out at 42 mph.) The average US price for a gasoline-powered car with all these capabilities in 2022 was $46,290—which easily exceeded that same year's median annual personal income for Americans after taxes. The improved Model T of 1925 sold new for $260—$4,533 in 2023 money, which means the best car for the average person a century ago cost less than a tenth of what we pay for the average car today in constant dollars. Somehow we went from cars that began as cheap workhorses that did everything to cars that are pricey and deadly show ponies, an evolution that doesn't match up with other major purchases of technology throughout the century. A radio in the 1920s—the only streaming device available in the era—cost $35 then, which would be more than $600 today, much more than the current cost of radios—or streaming video devices, or many flat-screen TVs. The same is true of washing machines, which are half the price today in constant dollars, and vacuum cleaners, with highly rated models now a third of the cost of their ancestors. Better technology, better materials, and modern manufacturing processes have made most machines that consumers buy today both superior and lower priced over time. But our cars have gone the other way—big time.

That's why our cars are stupid. They are not designed for the way we actually use them, nor are they designed to be affordable necessities. Just look to annual data gathered by the US Department of Transportation, which tells us that the average American drives 13,476 miles a year. That works out to an average of about 37 miles a day for each of us. Urban drivers generally run well below that, while rural drivers tend to go farther—Wyomingites being the most extreme example with an average of 66 miles a day. Drivers in most other rural states average 40 miles a day or less.

The average American covers those miles with four separate trips a day, which averages out to just over 9 miles a trip, with a trip defined as going to a place where you stay ten minutes or more before leaving on your next trip. We have all that horsepower in our cars, all that capacity, all that range, and we use our four-ton, $46,000 vehicles to go 9 miles, most often with only one human in the car.

It's the "one person and a quart of yogurt" problem, says Canadian energy expert Kate Daley—although for Americans, it's probably more apt to say the "one person and a fast-food value-meal problem." Using a full-size car for lots of little trips, like for that yogurt or a quart of milk or a bag of burgers, makes no sense, Daley says. "It's energy inefficient. . . . It's expensive."

Sure enough, the travel survey tells us that shopping and other routine errands account for the lion's share of our daily trips: 45 percent of them. Next come social and recreational trips, 27 percent (mainly visiting friends and family), and then commuting to work, which interestingly accounts for only 15 percent of trips. The average commute post-pandemic is 20.5 miles each way, and 76 percent of commuters make that drive alone.

Most of our trips are shorter than that, with half topping out at 3 miles or less. Ninety-three percent of our trips are 25 miles or less. The average person takes a trip that's 50 to 100 miles less than once a month. And the 500-mile road trip everyone frets about, particularly when considering an electric car that might need to be charged a time or two on such a trip: This is the main fear that drives people away from electric cars. Yet for the average American, more than a year will go by without ever taking anything close to that long a trip.

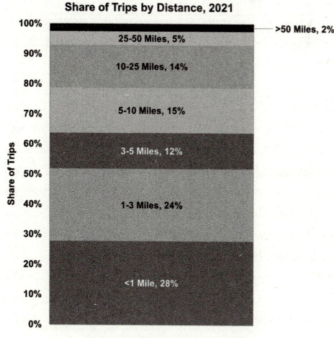

Share of Trips by Distance, 2021

Source: US Department of Transportation

You can see where this is going. Most of the time, our wasteful cars are a complete waste. They are the wrong tool for the job, clumsy and over the top. For almost all of our trips, we need the vehicular equivalent of a nimble tack hammer—a slim, lightweight tool useful for hanging picture hooks and other short and sweet tasks. We don't say, *Let's buy a sledgehammer for all our needs*, on the off chance that every once in a while we'll need to hit something really big and knock it down, which a tack hammer won't do. And if the sledgehammer damages the wall when we hang a picture, well, tough. At least we have one tool to rule them all, even if it is terrible at what we need it to do most.

That would be stupid. Nobody would do that. You want the tool to fit the job you most often need to do. But our cars are sledgehammers. All of them. Habit and that's-how-we've-always-done-it-ism keep us from seeing how terrible that is, or what a good solution might look like. We are spending huge amounts of money on cars exclusively designed in every meaningful way to solve a problem we rarely have, to drive distances that, on average, we rarely travel. What we don't focus on is what sort of car would work best for the most common driving scenarios, which is you and that proverbial container of yogurt alone in the car for 7 miles.

Electric Cars, +/-

Now the United States, under President Joe Biden, has launched a major national effort to fix our cars by electrifying the waste out of them. Billions are being spent on consumer rebates for new electric car purchases, along with massive corporate incentives to build new electric car plants and giant battery factories in America, instead of in the other countries that currently dominate the space—mostly China. The new symbol of clean, less-wasteful, less-climate-devastating transportation of the future is the electric car—the full-size, bigger, weightier, and costlier electric car. Yes, they're better than the status quo. Yes, they are much more energy efficient, and yes, the long-term cost of ownership is less than that of a fossil fuel car. They will cut pollution, including the heat-trapping kind, with a caveat: the climate benefit is less than it could be because our power plants are still so reliant on fossil fuels—unless you have a solar-powered charger or happen to live in a state like Washington or Iowa, where renewable electricity is the rule, not the exception.

But EVs are not really as revolutionary as they are being imagined and manufactured these days. We're just putting batteries in the same old sledgehammers—and jacking up the price. In 2023, that meant $55,000 on average to go electric.

The main reason EVs are so expensive—besides carmakers wanting to profit from premium rather than basic cars—is that we have been convinced to be obsessed with the range of electric cars. They have to go 300 miles on one charge, or 400, or more. Anything else will create something called "range anxiety," we are told. The national media is obsessed with this, with story after story detailing how someone or other who buys an electric car sets out to drive cross-country, then is dismayed when they "discover" that it can be hard to find places to charge along the way in the still early-times system of EV charging. This is depicted as *the* big problem with electric cars, and then that becomes the public perception—even though we know from data on how we actually drive that this is a made-up problem. Most of us rarely, if ever, make long-enough trips for that to matter. There are other ways of tackling the long-trip scenario without shelling out Americans' annual median take-home pay for a battery-powered sledgehammer on wheels.

We need to rethink what a car is and what we should pay for it. And we don't need to wait for car companies to tell us what we need or want. We can tell them. And actually show them. This can happen quickly, easily, and fairly cheaply. It doesn't require everyone to get a $55,000 electric car they can't afford. Which is a good thing, because that is the path to failure. We don't even need to get rid of all our gas-burning cars in order to make huge progress on cutting waste, pollution, car crashes, and climate disruption while also saving big bucks. We just have to stop using those sledgehammers so much.

Peachtree Perfection

This is why Peachtree City is important. We have to think like Peachtree residents. Because they're the ones who are making sense. We need the golf cart. It's the cheap and easy solution staring us all in the face: street-legal golf carts and electric bikes for most of our needs; regular cars on those rare days you need them—full-size electric ones if possible. But the greenest car is the one you don't use. The markets are exploding for both street-legal golf carts and e-bikes, which are the most popular EVs in the country. You see them everywhere. Do 90 percent of your trips in one of those small "personal" EVs. Keep your sledgehammer EV or even fossil fuel vehicle in reserve for those rare long trips. Or borrow or rent a big, long-range vehicle when you really need it for those few long trips and forget about electric range anxiety—and big car costs. Regardless, your transportation costs and waste footprint will drop sharply, your risk of being in a serious crash will plummet, and—just ask Mayor Learnard over in Peachtree City about this part—you'll be way happier.

As the mayor takes me on the grand electric tour of Peachtree City, most of the other NEVs that pass going the other way are fancy, sleek, and smooth-riding wonders, tricked out with sound systems and heaters and various other bells and whistles. Learnard's, on the other hand, is a decidedly low-tech golf cart, shaking and bouncing like a shopping cart on cobblestones. "I've got a horn, though," she volunteers, and gives it a couple of taps to prove it. But the sound is more tricycle squeeze-bulb toot than a legitimate get-out-of-my-way full-metal honk, which just adds to the charm.

The whole effect of this engineer and college professor turned small-town mayor bouncing along a lush woodland path in a golf cart is like stumbling into an alternative universe designed by twenty-first-century hobbits. It's an enchanting blend of nature and human technology that fits rather than clashes. The reds, golds, and oranges of autumn are brilliant in the morning sun, and through them, the deeply blue lake can be seen, its surface dotted with ducks paddling and diving for lunch. A couple heading the other way in their blue electric cart smile and wave and exchange greetings with Learnard. Everyone in golf carts here does this. It's a little unnerving, really: commuters aren't usually so chipper, or ready to slow down or to stop to greet their mayor or ask a question about some tidbit of city business or local gossip. Happy commuters are an oxymoron every other place I've ever been, but apparently not on this city's network of carless pathways.

About those pathways: they are the secret sauce of the city's success. This network of mini-roads winds through the town's five neighborhoods—or villages, as the city's designers called them back in the 1960s. These 100 miles of scenic mini-roads—officially designated multiuse pathways (MUPs), though no one calls them this, Learnard assures me—connect every major part of the city and several important adjacent destinations, including an industrial park and regional airport (the fourth busiest in the state). You can golf cart anywhere.

It was possible to create this unique road system because Peachtree City is much newer than most other towns in Georgia, chartered in 1959. It was laid out in what was then a modern, decentralized design with lots of green space built in, a ring of outer villages surrounding a central area dominated by Lake Peachtree and the civic center, with a spacious modernistic library as its signature element. The golf cart road system was not part of the original plan, but three golf courses were built, ringing the community. In the town's early days, long before its villages were built out, residents ended up cutting trails through undeveloped fields with their golf carts as they traveled between home and golf course. Eventually the town leadership saw the potential and began paving the trails and constructing that signature element of the town, its second road system. It was originally for pedestrians and bicyclists, but the golf carters started using it, too. The pathways eventually became a formal part of the city plan, and as new subdivisions were built, developers were required to tie into and expand the system so that a resident could get anywhere in Peachtree without taking a car or using the primary highways.

"And the rest is history," says Learnard. About four years ago, the city formally acknowledged its evolution by changing the city emblem from a golf cart with golf clubs in the back to a golf cart with no clubs. "It reflects that we are less a golf community and more a golf *cart* community."

Of the thirteen thousand households here, more than eleven thousand have golf carts. Many Peachtree residents commute to work in their golf carts. The local pizza place (where half the high schoolers in town seek their first job waiting tables), the shopping mall, the nearby Home Depot, and an array of restaurants, from southern cooking to Korean barbecue, are all accessible by golf cart. The local police force patrols in golf carts and ATVs. Everyone travels to the weekly farmers markets by golf cart, too. And there's a thriving market for used golf carts here. In her twenty-three years in Peachtree, half of them as a council member and now as mayor, Learnard has bought three of them used, and never paid more than $3,500.

A major rite of passage in Peachtree arrives for locals at age fifteen, when it becomes legal for an unaccompanied teenager to drive a golf cart to McIntosh High School. The parking lot there is crammed with hundreds of golf carts tricked out in all shapes and colors, with weatherized shells and decorated with the high school mascot or the occasional stuffed animal. The white lines in the parking lot are all laid out with golf cart–size spaces.

Somewhere between 60 and 70 percent of residents have electric golf carts, but the rest have gas-powered versions. Leonard says the heavy charging blocks of many of the older battery-powered carts are hard for some of the town's senior citizens to lug around. "They'd rather fill their gas tank, and that lasts them the year," she says. "They think that's a pretty good deal."

This is the beauty of the Peachtree City model: even the fossil fuel version of these little golf carts are low-impact, as are ones like the mayor's, with cheap lead-acid batteries in them, because they are good enough to get the job done. "I charge it overnight and I have never run out during the day. Never."

These golf carts are keeping large fossil fuel–burning cars off the road. There are no traffic jams, and Peachtree has yet to have a fatal golf cart crash. Golf carts are far more sustainable as a primary transport for our many short trips—which are almost all of our trips—than using our regular gas-guzzlers to run to the store or drop the kids off at school. Are they the most elegant solution to bring us into a new age of clean, green transportation? No, they are inelegant low-end tech. But they are also fun. And they work. But no, they're not as good as cars equipped with breakthrough solid-state batteries that are affordable, light, chargeable in minutes instead of hours, and can deliver a 700-mile range or more. Someday we may have them. Toyota announced in 2023 that it was getting close. Solid-state batteries, as the name implies, get rid of the liquids that reside in all of our current car batteries, replacing them with solid blocks of electrolytes that pack a much greater energy wallop. But so far, they are limited to small-format uses—in watches and pacemakers, for instance. They are too expensive as big EV batteries, and remedying that is the hard part we're waiting for. The Peachtree City model is what we can do *today*, without waiting for the latest holy grail tech, because frankly, we have done enough of that kind of waiting already. We should embrace the good enough. Because our cars, as they exist now, really aren't good enough.

The EV Sweet Spot

And increasing numbers of people here and around the world are getting that. This kind of low-speed, short-range vehicle is catching on all over. They've long been a staple of university campuses, and now they can be seen tooling around California's beach towns, multiple Florida and Arizona communities, and South Carolina's Charleston and Hilton Head Island. In Myrtle Beach and most of coastal South Carolina, traveling by "golf car" is a way of life. The global market for NEVs is exploding, reaching almost $12 billion worldwide and expected to double by 2030. Meanwhile, the electric bike phenomenon is well documented and even bigger than NEVs, a mobility option favored by young singles and students because of the low price point (highly rated models start below $1,200). Annual US e-bike sales now exceed $1.3 billion and grew by 33 percent from 2021 to 2022. E-bikes are America's most popular EV, even if, in Peachtree City, the golf cart is still king. Jasmine Gonzalez is a typical LA e-bike rider, recently out of school, using her pedal-assisted electric bicycle, with its 60-mile range, to zip along in traffic or loading it on an LA Metro bus for long trips across the Los Angeles Basin, then riding the last mile or two to her destination. She travels that way to the Sowing Seeds of Change urban farm in Long Beach, California, to volunteer with the program there that trains young people to create their own home gardens. Gonzalez is a home vegetable farmer herself, and the sustainability of her electric bike makes her happy.

"I think this adoption of small EVs is the mobility solution we've been waiting for," she says. "Small electric vehicles do what we need done, at a much lower cost. I love mine! It's my baby."

Gonzalez is correct, and she and the rest of Gen Z are leading the charge, though the sales numbers show that many people across generations are catching on. But could it be accelerated even more? Could using a small EV for everyday trips become the dominant way of travel? Could America really downshift from its current infatuation with transportation sledgehammers and come in for a soft landing into the Peachtree City model?

Given that about six out of ten US households have at least two full-size cars, replacing one of them with an e-bike or a golf cart could be a no-brainer for many families sick of the high cost of fueling, insuring, and financing multiple cars. A good nudge in that direction would be if the federal government retooled its $7,500 EV tax credit incentives to stop prioritizing the purchase of high-priced electric sledgehammers, and instead encouraged Americans to drive like a Peachtree City high schooler or an LA Gen Zer. Right now these short-range mini-EVs—let's call them "alt EVs"—don't qualify for much, if anything, on the rebate front, though some states have e-bike incentives of their own. We should be doing everything possible to encourage use of these short-range alt EVs, because they have a far smaller pollution and climate impact than the use and manufacture of full-size EVs, and would ease demand on already-constrained supplies of the essential battery mineral lithium. And as we know from Peachtree City's experience, its alt-EV model does the next best thing to getting rid of fossil fuel cars—it keeps them parked most of the time. The gas-guzzler you don't use is even greener than the full-size EV you do.

A recent emissions study found that if we turned 15 percent of our short-range trips by car into e-bike trips, it would reduce total greenhouse gas emissions from cars by 12 percent. That's an incredible reduction—done not by replacing cars, but by using e-bikes to *reduce* car use. If we change our prevailing habit of thinking a car is good for all our trips and instead choose the option that best fits the task, we save money, energy, and wear and tear on our expensive motor vehicles. If you think this makes sense for you, embrace it. There's no better way to promote systemic change than embracing it ourselves.

I break it down this way: For a short trip to the local coffee shop, I walk. For picking up a bag of groceries at the supermarket, I'll ride my bike. If I still commuted to work, an e-bike would be ideal for that 10-mile jaunt (and I don't have a golf cart yet, but I'm envious of my neighbors who do). And for a long weekend up the coast with the dogs (two greyhounds and a collie), we take the car—though without dogs, the Amtrak train from LA to the Central Coast is a great option. It's all about choosing the least costly, most efficient and, for the most part, most pleasant transportation method. Having a full toolbox, rather than just a sledgehammer, makes it possible, and for many people, an alt EV is the answer.

But alt EVs don't work for everyone. Weather can make them challenging in some places (although there are a surprising number of cold-weather e-bikers and golf carters out there), and landscapes and travel distances matter. Not everyone is an average driver. That means there's a hole in the emerging electric vehicle market that must be bridged: the gap between the cheap alt EVs that can get you through a day of short trips and the expensive full-size EVs with long ranges that many don't need or can't afford at that $55,000 average price point. To truly shift away from the waste, pollution, and expense of fossil fuels anytime soon, we need an affordable "good enough" EV with a range at the midpoint between Mayor Learnard's golf cart and the current market leader, Tesla. Currently, the closest cars to that in the US are the Chevy Bolt, with a base price of $26,500 (and which General Motors canceled then uncanceled in 2023), and the next cheapest EV, the Nissan Leaf, at $28,140. But that's not good enough.

Filling this gap would require an electric car that fills the niche of the old 1968 VW Beetle, a basic, small, no-frills town car with modest comfort, modest power, a modest weight of 1,745 pounds, and a very modest price of $1,699 new ($14,896 in 2023 dollars). It was homely, dated, starkly simple with minimal creature comforts, yet the classic Beetle became the sixth-best-selling car in history (it was number one from the 1970s to the '90s, when the Toyota Corolla eclipsed it). Or consider my first new car, a 1981 Honda Civic hatchback: Much more comfortable than the Beetle, well-appointed but not the least bit luxurious, my Civic weighed even less than the Beetle, at 1,664 pounds, and sometimes got an astonishingly efficient 50 miles a gallon on the highway. And its base model price was remarkably similar to the Beetle's: $4,568 ($15,333 in 2023 dollars).

We built light, basic cars with mass appeal back then. They were affordable, they got the job done, and people loved them. The Civic is on the list of the top five best-selling cars in history, ahead of the Beetle. It's not that US carmakers can't make good cheap cars anymore. They just don't like doing it, because the premium vehicle segment, and the obese SUVs they relentlessly market, have higher profit margins. China, where one-third of new car sales are electric, sells EVs equivalent to the Chevy Bolt for about $10,000—less than half the Bolt's price tag. And there are much cheaper EVs available there: General Motors is in a joint venture with China's top automaker, SAIC Motor, to build subcompact electric commuter cars sold in China for $4,500 ($5,000 with air-conditioning). So there is no reason why we can't electrify a car for US consumption as good as my old Civic today at that same price point, cutting costs by giving it a single-charge range roughly equal to three times our daily 37-mile drive.

That would be a car that could go 100 miles, with an option to upgrade to 150 miles. Such a car would meet the needs of most Americans almost all the time. An alt EV for everyday use and a just-good-enough EV as a household's second car is the imperfect but pretty damn good solution to all of our big crises—energy, environment, climate, cost of living.

To get there, federal incentives to both manufacturers and car buyers must be shifted away from the premium rich person's EV now dominating the market to e-bikes, golf carts, and some version of a dirt cheap e-Beetle. A price tag below $20,000 would be okay, but the ideal should be closer to the Beetle-Civic price point of $15,000. The industry won't do it on its own, but those hundreds of billions in federal incentives could, would, and should be used to channel all that genius innovation into building the cars that Americans actually need and could afford.

And for doubters that this is possible, a little history. In 1907, early electric cars were vying for dominance with early fossil fuel cars. Electrics had set all the speed records, had won the first race on a track, beating out the fossil fuel competition. And they were reliable, quiet, and clean, compared with noisy, smelly, and mechanically cranky early gas cars. The leading EV of the era was the Detroit Electric, which could reliably go 80 miles on a single charge—well over a hundred years ago. It was tiny and had the elegant look of Cinderella's carriage. Steered by a tiller like a boat's, not a wheel, it was built as a town car, limited by the weight of its primitive batteries to around 20 miles an hour for city driving. Sound familiar? Yes, the Detroit was the first NEV—the neighborhood electric vehicle, the street-legal golf cart before there were golf carts—with an awesome range and optional range-extending battery upgrade of special nickel-iron batteries from the genius inventor of the age, Thomas Alva Edison.

And one more bit of more recent history. In the 1990s, California imposed what turned out to be a short-lived electric car mandate that sought to force automakers to bring zero-emissions cars to market. The state did this after General Motors showed off a prototype of a car that would eventually be called the EV1, which essentially replicated the performance

Source: saturdayeveningpost.com

of the Detroit Electric but with the normal high speeds expected of a modern car. It got mileage of 80 to 100 miles with lead-acid batteries— advanced versions of the Detroit Electric's batteries, and similar to the batteries that run the golf cart driven by the mayor of Peachtree City.

With the computer management of car power systems available now, and better manufacturing techniques for those cheap lead-acid batteries that need none of the rare substances used in Tesla-class electric cars, a latter-day EV1 could be rebooted to become the good-enough, cheap-enough car to drive fast and promote widespread adoption of supercheap electric cars. If it could be done in the early 1900s—and there are still Detroit Electrics tooling around, and the videos are on the web to prove it—and done again in the 1990s, we can do it now, and do it much more cheaply than any EV currently for sale. The good-enough EV with cheapo batteries that require no lithium, no nickel, and no rare, expensive elements comes with an added feature: we already have a robust and highly successful recycling infrastructure for the lead-acid batteries we use to start every fossil fuel–powered car in the world. It's the no-brainer solution no one is talking about, in part because of the phony range anxiety issue, in part because American car manufacturers don't really want to build the car we need because cheap cars don't make them as much money, and because, to get back to my central premise, we have a long history of being stupid about our cars. But this is the solution we can have now, at a time when waiting for lower emissions is just not an option.

Convincing Americans to go back to the past and embrace small, light, and simple once again, to let go of the false narrative about range anxiety, and to garage their gas-guzzlers and SUVs except for that once-a-year long road trip is going to be the harder mountain to climb. Will an irresistibly low price point and the well-documented Gen Z interest in living more sustainably make a difference? I don't know. But I'd buy that e-Beetle in a heartbeat if it ever gets made.

Even without that gap bridged, using the e-bike and the golf cart as our everyday transportation is a viable solution, too. Electric golf carts work on city streets just fine—the folks in Peachtree City leave their special paths sometimes, as drivers in other towns with street-legal golf carts do all the time. Kim Learnard is all in on this trend because she is an environmentalist and believes that this model, with or without the unique pathway system of Peachtree City, can help save the world. But not everyone in Peachtree City sees that as a motivation, and they don't have to. They love the economics of not having to fill their gas tank up every week, of not having to fight traffic, of enjoying the view as they trundle along Peachtree City–style. The lifestyle—the ease, the social part that you lose sealed up in a car zooming along at 65 miles an hour—is the hook everyone can get on board with.

Peachtree City has one particularly lovely tradition that capitalizes on this, an informal gathering called the Sunset Club. There's a spot along Lake Peachtree where there's room to pull off the no-cars-allowed pathway and park, where a perfect view of the sun as it sets over the lake awaits. Friends and neighbors glide up in ones and twos as sundown nears, a chance to catch up and spend the end of the day together, sharing homemade snacks and cool drinks.

"Sometimes," Mayor Learnard says, "a dude with bagpipes shows up, hops out of his golf cart, and starts playing." It's one of her favorite Peachtree quirks. "It just makes me happy," she says. "What could be stranger or more wonderful than that?"

10

The Wallet Ballot

The first impression when you walk into GoGo is the lightness—the warmly lit space, the clean blond-wood countertops, shelves, and displays, the artful arrangement of carefully curated wares. Think Apple Store meets old-school general store. The focal point is the long main counter fashioned from salvaged wood and, behind it, what looks at first glance to be a display of exotic ice creams with amusing flavor names: Lucy in the Sky, Blackberry Betty, Johnny's Cash, Sexy Sadie. Owner Laura Marston opens up one dark concoction called Super Star, releasing the subtle, spicy aroma of cardamom.

On second glance, it's clear this is not a dessert bar but a colorful arrangement of some sort of personal care product in bulk containers.

"Natural deodorants," Marston says. "One of our most popular items."

Her shop's full name is GoGo Refill, where such products as those all-natural deodorants from a Canadian company called Routine go home in your own reusable containers, or one you purchase there for future reuse. The shop's collection of products for the home, kitchen, bath, and personal care at first seem typical of a Target or Meijer or Fred Meyer or HomeGoods, but they stand out for their sustainability and their lack of disposable plastic containers, packaging, and unhealthy ingredients.

That means no more plastic waste in the form of bottles for shampoo, conditioner, hand soap, body wash, dish soap, hand sanitizer, dishwasher detergent, hand and body lotion, toothpaste, mouthwash, linen spray, carpet deodorizer, all-purpose counter cleaner, or all-purpose surface and glass cleaner. No more fossil fuel–derived chemicals in your shampoo or toothpaste, no hormone disruptors or carcinogens in fragrance formulas— "trade secret" ingredients that, legally, don't have to be on the label but that labs have sniffed out in many common products.

Instead, just bring your own containers—glass, metal, even reuse your old shampoo or other plastic bottles—and fill them up. The deodorant, made with proven-safe minerals (baking soda or magnesium hydroxide) and plant-based materials, has the consistency of a soft clay. Just scoop a pea-size dollop out with a finger and rub it into your pit. Yes, it's basic, but that means no plastic dispensers, no roll-ons, no little gears and wheels to turn to push up the stick of deodorant—none of which gets recycled. A simple, effective, safe, zero-waste stink stopper with nothing to throw away when it's used up.

Such things add up. The average American consumer spends $400 to $500 a year on household and personal care products, representing more than a hundred disposable plastic bottles and containers landing in the trash per household. We have plastic forests of those bottles under the kitchen sink, in the bathroom, on our shelves. Nationwide, that's a lot of disposable containers. Your personal contribution to the flow stops if you patronize one of the growing army of zero-waste and refill stores around the country—currently numbering about 1,400 in a conservative count. They're in every state in the country, and in thirty-two states they number in the double digits. California has over 150. These are in addition to a major source of bulk groceries—the country's 325 customer-owned food cooperatives.

Oh, and you won't pay a fortune for the privilege of zeroing out your waste from these product lines: GoGo Refill's bulk prices are competitive with the per-ounce prices of products' name-brand, plastic-encased equivalents. The Wicked Strong dish soap, made from recycled cooking oil, is only sixteen cents an ounce, versus the typical supermarket price of twenty to thirty cents an ounce for name-brand dish liquids.

When she opened GoGo in South Portland, Maine, in 2019, Marston

was new to the refill universe and had never run a brick-and-mortar business of any kind before. She had worked as a software strategy consultant
for business start-ups, while raising two kids with her husband, a garage-
door wholesaler. Launching a retail business of her own wasn't on her
radar screen.

"I was working for other people who were starting their businesses,
helping them turn their dreams and interests into software products. And
as I went through that process with them, I started thinking: *What would
I do if I could turn something I'm deeply interested in into a business? What
would that look like?*"

It so happened that right around then, she had been making a con-
certed effort at getting plastic waste and single-use packaging out of their
home and lives. She also had become passionate about finding healthier
products without all the concerning chemical ingredients that often come
in the personal care products that are encased in all that plastic. At first,
she went to her comfort zone and thought about how she could build a
software solution for people who wanted to ditch plastic and how that
might work. But that didn't really appeal to her, and she began thinking
that a more hands-on, community-oriented approach was what she longed
for. Instead of an app for that, she wanted a store for that, in the town she
loved, with its long-standing commitment to conservation, but that was
stuck with an increasingly difficult recycling system. She knew of no store
in the area that specialized in the sustainable, low-waste products that
many in South Portland wanted.

"And that's when the idea for GoGo was born."

The name took longer than the business plan to get right. She wanted
something that would counter the notion that shopping zero-waste was
hard or inconvenient, a name that suggested shopping that could be a
quick in-and-out experience for people. "GoGo" finally came to mind,
and it stuck. It was good messaging, even though the vast majority of customers take time to browse to see what new products Marston is testing
out in the store, from the steel-mesh pot scrubbers that last for years (instead of the ephemeral steel wool sold in supermarkets) to the recycled

paperless paper towels that people snap up. She tried a lot of toothpaste tablets, natural deodorants, detergents, and shampoos before settling on the ones she offers customers. Marston and her young sales staff test out everything GoGo sells, she says, so they're not just serving up marketing information when customers ask questions.

A little more than a month after launch, her business hit the roadblock that shook the world, COVID-19, and it looked like her dream might come to a short but not sweet end. But her customers—existing and new—quickly tired of the endless Amazon boxes piling up in their basements and trash cans, and they rallied to keep GoGo going as a curbside business. Now with commerce back to normal, business is brisk and profitable enough that Marston has opened a second location 27 miles away in Brunswick, Maine, with a bright red GoGo van for pop-ups—a mobile refill station—at events and farmers markets as well.

The months as a curbside-only business taught Marston a few lessons, the biggest one being that customers like GoGo to fill up the bottles for them, rather than the original model of self-service pumping. Now they bring in their empties, put in their order, and then go browse. Meanwhile, the staff dispenses shampoo, detergent, and other products out of beer keg–size white drums with pumps and taps on them that make refilling quick and painless.

The other post-pandemic surprise was the demographics of her customers. She expected her mainstay shopper would be people like her: working parents in their thirties and forties.

"But no, from day one of reopening our doors, it was the eco-grandma. That's what I call her anyway. Oh, I knew she would be one of my customers. Something we do in software design is we'd really flesh out the personas of who we are building things for, so I did that for GoGo. I just had no idea the eco-grandma in her sixties and seventies would be our biggest customer group. But they come in and look at the natural products, and the vinegar-based cleaners with no chemicals, and their attitude is *Oh, I've been doing this since the sixties. I've been an environmentalist the*

whole time and just hacked the system myself before you got here. But this is a
lot easier!"

Another set of early enthusiastic customers were staff members from
the Natural Resources Council of Maine, which is how she met Sarah
Nichols, who was still trying to push through the state's breakthrough
extended producer responsibility law for packaging and containers. Marston immediately volunteered to help and eventually testified before the
legislature as one of numerous local business owners who supported the
new polluters-pay law. She also was instrumental in the state's decision to
change rules that barred zero-waste stores from selling bulk groceries to
customers who bring their own containers. It turned out to be much easier
to be heard by government, at least in Maine, than she dreamed possible.
Her investment of time and effort wasn't in vain, drowned out by the lob-
bying and weight of powerful corporations invested in keeping a disposable economy. The chorus of small-business owners and citizens won the
day. That feeling of futility that keeps people from taking this sort of action is what the other side depends on, Marston says. It worked on her in
the past, but no longer. She had become one of Nichols's Marges.

"I feel so strongly that solutions to our environmental crisis can't come
from individual action alone. It can't be just the responsibility of the consumer. There has to be regulation, and we have to hold corporations responsible for their pollution and the waste that they create and the impact
they have on the climate. We have to have a multipronged solution."

Laura Marston's decision to ditch plastic ended up becoming a waste-
fighting triple play. She reduced her and her family's personal waste footprint. She helped others do the same through her business. And she took
a leadership role in the adoption of a statewide solution to plastic waste
that compels its creators to clean up their own mess. Marston had no special training in any of this. She learned as she went, got help from a local
nonprofit and expert who helped her navigate the citizen participation
piece of our democratic process, got her facts straight, and in the end felt
she made a difference. Now she is leading an ad hoc group of Portland

residents and businesspeople crafting a plan to create a reusable container system for the city. "It's the next step," Marston says. "I'm excited."

Her example is as clear a demonstration of citizen-consumer power as you could ask for. The causes we support with our voices and our votes matter. Our choices matter. The way we shop matters. It's never too late to alter the total-garbage trajectory—and to enjoy the environmental, health, and economic benefits those choices bring.

I asked the owner and founder of GoGo Refill to pass on her top tips on low-waste living and shopping. Here they are, Laura Marston's personal approaches to fixing waste:

1. **Begin with the end in mind.** Thinking about what will happen to a product or package at the end of its useful life before you buy it is key! That will help you say "No thank you" to wasteful products.

2. **Reuse, reuse, reuse!** Beyond your reusable tote bags, focus on ditching disposables! Replace disposable items in your home and life with reusables. Start simply, with a water bottle and coffee mug. When you're ready for the next level, try replacing paper towels, coffee filters, and plastic baggies.

3. **Package-free is the best** . . . but some packages are better than others. Buying package-free from a co-op or refill shop is great. Reusing a container that you already have even just one time cuts its carbon footprint in half. We are surrounded by durable containers in our homes and in our lives, and just reusing what we have reduces the size of our footprint.

4. **Make yourself a do-not-buy list.** This is great for the grocery store. Decide ahead of time what you won't buy—it makes real-time decision-making much easier. On my "do not buy" list: berries when they're not in season, hummus or salsa in plastic containers, meat

from the grocery store. This looks different for everyone, so figure out what works for you.

5. **Don't buy anything!** The least wasteful thing to use is the thing you already own! It might not be sexy or cool or new, but using up what you have before replacing it is the way to go.

6. **Engage in the secondhand and borrowing economies.** Borrowing, trading, and buying secondhand are becoming more popular through things like Buy Nothing groups, Facebook Marketplace, and Craigslist. All of these alternative ways of consuming save money and take things out of the waste stream, and together we're walking away from the notion that we need to be overconsuming and overproducing all the time.

The Monster in the Closet

Marston's strategies are all aimed at moving us away from disposable, single-use items and toward reusing (or selling, trading, or giving away) what we already have, and buying used or refurbished products. When you do buy new, shop for quality and cost of ownership rather than low price and low quality. And when something breaks, don't throw it away. Fix it or find someone who can fix it for you. Clothes can be mended. Lamps rewired. Community repair cafés are launching in cities all over the country to help you get stuff fixed.

This way of consuming turns a linear buy-use-waste consumer economy into the more sustainable circular economy—giving new life to existing products instead of landfilling or trying, mostly unsuccessfully, to recycle them. Reuse radically reduces the environmental impact of manufacturing and transporting new stuff. It saves materials and fuel, and is an efficiency powerhouse. Used and refurbished electronics can be found at substantial discounts, as can furniture, major appliances, books—virtually any consumer good can be found secondhand.

The hottest market in the reuse economy—a major source of waste as well as an opportunity for reducing it—is something we use every day, all day, everywhere we go. It's primarily made out of plastic, so reusing it also reduces that pernicious pollutant, along with its contribution to our weekly credit card's worth of plastic ingestion.

This is, of course, the monster in our closets: our clothes. Fashion is the gateway where many people enter the reuse economy, and no sector needs a break from waste more. The global impact on the environment, climate, and plastic pollution makes fashion very ecologically unfashionable.

Fashion waste has been accelerating for years. Global emissions from the apparel industry are projected to increase by 50 percent by 2030 unless the industry and consumer behavior change. In an in-depth analysis and call to action a few years ago, the Ellen MacArthur Foundation summarized the unfolding disaster that is our fashion industry and its linear production process that takes, makes, and disposes all with very little recycling. It's a $1.7 trillion global business, and it was built in the modern age on a model of churning out cheap, mostly plastic clothes that are worn for only a short time, with almost half that $1.7 trillion value ending up in landfills or incinerators every year.

The heat-trapping pollutants generated by fashion production currently total 1.2 billion tons a year—nearly 10 percent of all greenhouse gas emissions worldwide. Fashion emissions churn out more climate-disrupting emissions than all international flights and all maritime shipping combined—including all of the world's 5,600 giant container ships, the most polluting vehicles ever conceived (and which, of course, transport virtually all clothes sold in the United States).

The chemically intensive process of making and dyeing textiles causes significant air and water pollution at the production sites, and raising the cattle that provides leather (as well as beef) is responsible for 80 percent of Amazon rainforest deforestation. Every seventeen pairs of leather boots require an acre of clear-cut rainforest to raise the cows (replacing the natural carbon-capturing trees with major carbon-emitting cows). Another acre is needed to make just ten leather bags, and nine leather jackets will kill every tree on two and a half acres. You can see how our shopping habits and choices at the end of the supply chain add up at the front end.

Basically, we buy a lot of clothes. And we throw away a lot of clothes—more than eighty pounds for every person in America ends up in the landfill every year. About 11 percent of US landfill contents are textiles, leather, and rubber. Worldwide, the clothing that gets incinerated or dumped each year is equal to one garbage truck full of fashion every second. All that throwing away means a lot more buying. In fact, we're buying twice the clothing we bought twenty years ago, but we are throwing it away twice as fast, too. We are paying less, but getting less bang for the bucks we are paying. This is the linear economy of waste at its worst.

Fast fashion—cheaply made and cheaply priced trendy clothing—has created apparel that is almost as disposable as plastic soda bottles. The irony is that a high proportion of fast fashion is made out of precisely the same plastic as a soda bottle—polyester, which is in 60 percent of everything you wear. Plastic bottles are more often recycled (when they are recycled at all) into polyester for clothes than they are turned into new bottles. This is a fake circular economy. In a real circular economy, materials would never be thrown away, but remanufactured into the same type of product continually. Instead, bottles are becoming clothes and, unfortunately, once in polyester form, the material is too compromised and contaminated with dyes, chemicals, and additives to be recycled conventionally into anything else. That's why so much fast fashion ends up in landfills and incinerators.

Quantity has been elevated over quality, but a marketing push to make clothes and styles much more ephemeral allowed the industry to present the rapid obsolescence of its creations—their poor construction and short life span—as a feature, not a bug. Fashion makers deliberately overproduce these flimsy rags, knowing that they'll be eclipsed by some new trend before they can sell out. Each season fast fashions that fall out of trend are trashed or burned—a shocking 30 to 40 percent of new fast fashions made. The business model is literally built to be garbage.

Cheap plastic clothes also shed fantastic amounts of microfibers as we wear them and particularly when we wash them. The average load of laundry sheds about 700,000 microfibers. Over the course of a year in the United States and Canada, clothes washing sheds 85 quadrillion microfibers, 3.5 quadrillion of which are believed to escape into waterways and the ocean. That's 878 tons—the weight of ten giant blue whales. Microfibers are tiny and almost weightless, so that many tons of them spread far and wide in aquatic environments—yet another source for that weekly plastic credit card we're all eating. The total global plastic microfiber pollution from laundry is estimated to be 10 percent to more than 35 percent of the total amount of plastic entering the ocean every year.

Only one country in the world, France, requires manufacturers to equip washing machines with filters that stop the fibers at their source. It is vital that this become universal, and that means public demand and government action are needed. California passed a washing machine law in 2023 that mirrors France's, but it won't take effect until 2029.

For now, you can reduce fiber shedding by washing full loads rather than small ones—less bouncing around inside the machine means fewer fibers are jarred loose. And front-loading machines that just tumble the clothes and use less water cause less wear, tear, and fiber shedding than top-loaders with agitators that collide much harder with the clothes inside. Consumers can install after-market filters on the washing machine drainage hose, but they are clunky add-ons, and a starter kit typically costs more than $60 plus shipping, with ongoing costs for replacement filters. Some people have taken on the effort and expense because they care deeply about plastic pollution, but this needs to be standard equipment on every machine. Microfiber shedding could be stopped by the washing machine industry, rather than their aiding and abetting it, and under the polluter-pays principle, the fix should be universal and installed at the factory as part of the machine's design. If the same extended producer responsibility principle used to compel makers of wasteful packaging to pay for cleaning up their mess was in place for washers, this would have been done a long time ago.

What Can We Do about Unrecyclable Clothes?

Until that happens, a more immediate solution is to buy fewer clothes and invest the money you save in high-quality items with higher natural fiber content. Such clothing tends to last longer, hold up better in the wash, and have less plastic to shed. And one of the best ways to do this also happens to be one of the fast-growing trends, particularly among millennials and Gen Z shoppers: thrifting for used clothes, which stops fashion waste in its tracks one piece at a time.

The resale business for clothes is projected to grow three times as fast in the next few years as the fast-fashion retail market. At least eight out of ten Americans buy or sell used products now, and young thrifters are driven as much by the sustainability aspects of the reuse economy as they are by its economic benefits, multiple consumer surveys have shown. The global market for used clothing is expected to grow by 127 percent, from $119 billion in 2022 to $218 billion in 2026—with North American sales accounting for $91 billion of that total, according to analysis from online thrift store thredUP.

Thrifting influencers are all over social media, thrifting clubs are making the reuse economy a social phenomenon, and a sector that not long ago carried a negative stigma is now a vibrant community-based shopping trend drawing more customers to thrift stores than shopping malls. Asia Marquis, of Dayton, Ohio, and her Thrift Sistas Club made national news in 2023 sharing tips on how their monthly thrifting posse uses fashion ideas from vintage TV series, Pinterest photos, magazine spreads, and other inspirations to assemble ideas for their wish-list approach of navigating the thrift world. When you have specific themes and types of outfits to search for, Marquis says, the sheer volume of goods and occasional disorganization of thrift stores doesn't become overwhelming. Two other Thrift Sistas pro tips: Mondays and Tuesdays are good days to thrift, because most donations come in on the weekend and you can get first crack at that week's new stuff. Also, shop out of season to avoid competing with the other customers: fewer people will be looking for winter coats in July, while shorts and tank tops will be all yours to choose from in winter.

The age-old art of mending worn and torn clothes is making a big comeback, too, on social media and in the pandemic-fueled explosion in sewing machine sales. A group of young menders meets regularly to fix their duds and to teach others to do the same at the University of Minnesota Twin Cities campus, a trend that is catching on at other schools, powered by its economic, sustainability, and social benefits. It also extends the life of beloved and well-worn clothes. Patagonia, a leader in embracing corporate product stewardship, has its ReCrafted and Worn Wear programs to repair and resell used clothes or, if they are too damaged, to bale them up and ship them to Suay Sew Shop in Los Angeles for reuse and reinvention.

The mend, reuse, and renew principles of a true circular economy all intersect at Suay. According to Patagonia, its 5,000-square-foot shop and retail space near the Los Angeles River asks and answers the question "What if we could wear our garbage?"

The company, founded in 2017 as a community-based center for sustainable fashion and sustainable jobs for thirty workers, offers a living wage in an industry known for poor pay and worse working conditions. Suay upcycles fashion waste into new products, diverting and repurposing nearly four hundred tons of discarded garments that would have been landfilled. The material from Patagonia and other sources gets sorted by fabric type and how various pieces might be repurposed, with some material landing in the "pocket box," other pieces into the "sleeve box," and on through every component of apparel and accessory. A former farmer, punker, and inveterate thrifter since seventh grade, Suay founder Lindsay Rose Medoff says, "We're actually doing the impossible."

Or rather, what she had been told was impossible.

Suay turns these trashed textiles into pillows, towels, bedding, totes and other bags, and clothing and accessories of all types: jackets, pants, aprons, sweats, cardigans, hats—a constantly changing array based on what sorts of materials come in. The pieces look like nothing else, with fabric and other materials mixed together into practical, wearable art— their origins as material from something else often clear to see and yet beautiful in their distinctiveness.

Fusing both the reuse and mending trends, Suay's staff hope others will see their business as a model for circular fashion that is anything but fast. If all the garments now being destroyed were handled in this way, the entire world population could be outfitted without manufacturing a single new article of clothing. Even achieving a fraction of that would be an enormous blow against waste.

The name of the company is fitting. It's a word that the owner of Suay had heard many times from her mentor, Tina Dosewell, a Thai immigrant and master seamstress who shared her knowledge and ingenuity in building Suay's business. The word is her callout when a remade garment is sewed and the work is finished to Dosewell's standard. That is when she says, "Suay!"

It is the phonetic spelling of the Thai word for "beautiful!"

Your Choice

Starting with a few simple, straightforward, different daily choices is the best path for individuals and communities to become more sustainable consumers, according to Sabrina Pare, the Detroit-based cofounder of the Eco-Tok Collective of seventeen influencers on TikTok (where she can be found as @Sabrina.Sustainable.Life). Pare posts about the daily choices that make her happy and make her life more sustainable, and she passes on her tips and product recommendations to her followers. Her message: "You don't need to be a scientist or work for an environmental organization to make a difference in the climate space. Every voice matters."

Pare's voice reaches 230,000 followers as she navigates a more sustainable life. She's a dedicated home vegetable gardener—her TikTok feed has lots of short, helpful videos on this subject, from natural pest-control methods (castile soap mixed with water in a reusable spray bottle) to composting with worm help (the worms do the work so you don't have to). For what she's not growing, Pare orders grocery boxes from Imperfect Foods, which sells cosmetically blemished but otherwise nutritious and high-quality produce and other foods supermarkets typically reject—the "ugly fruit" category. Ugly comes cheap, too. (Imperfect has merged with former competitor Misfits Market; together the companies have nearly a million online customers and have reported that they rescued about half a billion pounds of food that would otherwise have been food waste.)

In her posts, Pare likes to focus on small things that add up—such as her use of recycled printer paper and recycled/refillable ink cartridges in her home office, and she's especially adamant about the virtues of the recycled version of a product everyone needs but most don't think about: toilet paper. She's a big fan of ordering online the 100 percent recycled toilet paper made by Who Gives a Crap—especially its Happy TP line, which comes wrapped in beautiful art-print paper that she repurposes as gift wrap. This tends to be a bit more expensive per roll, but Pare points out how much good it does, based on data from the World Wildlife Fund, which shows that toilet paper made from virgin materials leads us to collectively flush the equivalent of twenty-seven thousand trees per day worldwide. Recycled content avoids that tree toll (so does bamboo-based toilet paper, as bamboo is a grass, not a tree, and grows fast). The reviewers at Wirecutter suggest the 100-percent recycled toilet paper made by Seventh Generation as a high quality choice that's price-competitive with regular TP.

Young influencers like Pare and others in the EcoTok Collective reflect a multigenerational movement toward sustainable lifestyles that is one of our most hopeful trends, from influencers who start as young as environmental activists like Greta Thunberg, who created a sensation at age fifteen when she started spending Fridays outside the Swedish Parliament demanding action against climate change, to then eighty-two-year-old James Cromwell, known for starring in the TV series *Succession* and for his iconic role as Farmer Hoggett in the classic film *Babe*. Cromwell superglued his hand to a Starbucks counter to protest the company's policy of charging extra for plant-based milk and was arrested in New York while protesting expansion of a natural gas storage facility related to fracking.

There are plenty of ways to have impact with less drama, of course. Pare's advice to people just getting started at becoming more sustainable is don't try to do too much at once. "Start slow, because it can definitely feel like you're taking on a lot at once. If you're able to connect with others who are passionate about it, it does make it a bit easier to get ideas and advice. Finding a community is super important if you're able to."

Purchasing Power

Pare's pursuit of sustainability began in 2017 when she read about the environmental, climate, and waste impacts of the factory farms that raise animals for food. She decided to switch to a plant-based diet and then began systematically looking for ways to be less wasteful and more sustainable in her choices, and she connected with a community with shared interests. This is a pretty common path, as our shopping choices are the low-hanging fruit of sustainability. It's something you can do now. That's why thrifting for clothes and other used products is so popular. That's where the multiple health, environmental, and economic benefits of whipping the waste can be both obvious and dramatic.

Shopping is one of those everyday activities that presents us with multiple daily choices that can be used to further invest in the total-garbage economy, or to steer Our Good Ship Consumer toward a more sustainable port of call. These can be big decisions: choosing to use your buying power for a heat pump or other Passive House systems, or for displacing some or all of your fossil fuel travel with an NEV or e-bike. Or, as many who embrace sustainability in daily life try to do, it can be an accumulation of small decisions that gradually replace the wasteful things in our lives that add up in terms of the environment and climate, and that, through their absence, also enhance daily life, health, and wealth. We have so intertwined our roles as citizens and consumers that we refer to the impact of these choices on markets and policies not just as our purchasing power but as "voting with our wallets."

It's a concept that is both troubling and empowering, and it's one that has often been deployed to stack the deck *against* sustainability. Companies have manipulated pricing of sustainable products they didn't really want to bring to market—from reusable beverage bottles on the low end to the remarkably good electric cars California briefly forced carmakers to produce in the 1990s on the high end. When they didn't sell because of uncompetitive pricing or other barriers (such as ridiculously difficult lease applications and waitlists), the industries invested in plastic and fossil fuel dependence had their *I told you so* moment. They argued that consumers voted with their wallets against the environmentally better options. Renewable energy opponents (fossil fuel companies and their front groups) were always quick to make this case in the early days of rooftop solar, arguing that federal subsidies that made solar more affordable for consumers amounted to the government "playing favorites" with market forces and violating the rules of the vote-with-your-wallet game. This argument was a hallmark of the Ronald Reagan White House's environmental policies as he slashed Jimmy Carter's renewable energy programs into irrelevance, setting back solar energy development by more than a decade while doing everything possible to ensure continued fossil fuel dominance. It was this radical change in policy by a president who told Americans that "trees cause more pollution than automobiles" that helped convince Exxon and other fossil fuel companies to abandon their research into climate change—which they knew was coming as far back as the 1970s—and to cancel massive investments in renewable energy research. Had Carter won a second term and kept the United States on course for a renewable future, the oil industry was expecting tough new fuel regulations, and the renewable energy research investments were their plan B. But Reagan won big, and Exxon stopped researching how to fix climate change and started investing in climate change denial.

This was perhaps the most disingenuous use of the "Americans voted with their wallets" argument in history, given that global fossil fuel subsidies have long been greater by far than any renewable subsidies. In 2022, the International Energy Agency estimated fossil fuel subsidies at an astonishing $1 trillion annually worldwide—for an industry with record-breaking profits that same year.

But the voting-with-your-wallet meme doesn't have to be weaponized in favor of waste and pollution—the concept cuts both ways. We can, in fact, use it to bend big companies and markets to our will by collectively voting with our wallets for low-waste and sustainable solutions—particularly when whole communities, campuses, and large public agencies do so in concert. Purchasing power is, simply put, power. The way we shop affects our personal quality of life, but it can also have broader influence at the same time. It's a tool in the fight against total garbage.

WHAT YOU CAN DO

- **Shop for zero-waste packaging.** That could mean patronizing a local refill store like GoGo, or a grocery co-op, or another market that sells bulk items, or a farmers market, where fresh food is available without packaging. It can mean insisting local restaurants and other businesses use sustainable takeout containers, utensils, and other materials, or you take your business elsewhere to a place that does. Or it could mean choosing online retailers that ship with sustainable and reused materials.

- **Consolidate shopping-related travel to avoid waste.** Every purchase is a trip, either in the form of a trip to a brick-and-mortar location or by ordering online—which is basically ordering a truck to make a trip to your house. Getting to a store by foot, bike, or electric power makes brick-and-mortar shopping the most sustainable choice in terms of a travel footprint. If you're going to drive a gasoline-powered vehicle to shop, then online shopping with multiple purchases in one shipment (rather than many one-item purchases) is less wasteful. A single truck making seventy deliveries in your neighborhood is more efficient than one person in one car fetching stuff for one household. If your local shopping place happens to also be a drop-off point for Amazon and you shop with the online giant, combining your local shopping with picking up your online goods is a sustainability home run in terms of transportation-waste avoidance.

- **Shop less.** The greenest product is the one you decide you don't need to buy after all. One popular tactic is a self-imposed wait-overnight rule for significant purchases. Leave it at the store or leave it in your virtual cart online, and see if, twenty-four hours later, it's still a must-have purchase. Half the time, it won't be. When you do decide to buy, consider local stores that carry zero-waste, sustainable, and earth-friendly products, or shop at one of the online retailers that specialize in green retailing, such as Thrive Market (groceries), Pact (organic cotton fashion basics), Package Free (sustainable personal care items), or the "Amazon of sustainable shopping," EarthHero. These eco-friendly companies are all highly rated by Treehugger.

- **Buy used rather than new whenever possible.** And sell, trade via Buy Nothing and similar community groups, or donate the stuff you no longer need or want. This is the sustainable circular economy in its purest form—giving new life to existing products and reducing the environmental impact of manufacturing and transporting new stuff. Used and refurbished electronics can be found at substantial discounts, as can furniture, major appliances, books (especially textbooks), and, of course, apparel.
- **Find a zero-waste or refill store near you at litterless.com.**
- **Find a repair cafe at repaircafe.org.**

11

Schooled

Whether the 130-acre campus is sunlit, snowbound, or glowing beneath the crackling curtains of the northern lights, the main attention-getters at the University of Minnesota Morris are always the same: Bert and Ernie.

Size has a way of standing out in this broad expanse of midwestern farmland and lakes, and this particular pair is *huge*. Bert, of course, is the taller of the duo, standing at 262 feet, just a skosh less than the Statue of Liberty, while Ernie comes up 30 feet shorter. Both, however, have arms as long as two big-rig trucks.

Bert and Ernie are a pair of 1.65-megawatt wind turbines—the nicknames are unofficial if permanent, cooked up by a recent Morris graduate, Sydney Bauer. The slender white towers are each topped with three whirling blades that create the illusion of sedate motion even as their tips exceed 100 miles per hour on this windswept campus. Together they kick out twice as much electricity a year as the campus actually needs—generating more renewable energy per student than any other university in the country. So Bert and Ernie also supply clean energy for sustainable farming research on an adjacent campus, and send the rest into the grid for surrounding communities. But these massive Muppet namesakes are only the most visible component of a far-reaching sustainability initiative here in the heartland, a remarkable partnership between a state university, a conservative rural city and county, corn and dairy farmers, and the 1,500

Gen Z college students who come here for a liberal arts education. It's the ultimate example of the economic benefits of sustainability and waste-cutting choices that also happen to be good for the environment and climate. "We never made it about climate," says recently retired city manager Blaine Hill. "We just did it because it makes sense. And the more we did, the more we wanted to do."

If you want to see what a community committed to tackling waste in all its forms looks like—a farm town with electric buses for public transit, students manning a countywide composting operation, cows grazing next to elevated solar panels that provide them with shade when it gets too hot, windmills powering the manufacture of fossil fuel–free fertilizer for local agriculture—Morris, Minnesota, is the place to be. Indeed, what's happening here is an environmental and sustainability partnership called the Morris Model, with its plan for energy storage and local resiliency, zero landfilling by 2025, groundbreaking research on sustainable farming, and 80 percent locally produced renewable energy by 2030. Even the city-owned local liquor store (it's a Minnesota thing) is solar-powered in this beer-loving part of the world and has the motto "We chill our beer with the sun."

Schools and Students Are Leading the Charge

Morris is not alone. Colleges, universities, and public school districts throughout the country are living laboratories of waste reduction and clean energy. They're training the next generation to live much more sustainably than their predecessors—and the students are not just along for the ride. Students are driving this train, prodding and insisting that their schools start changing the world now, starting right in their own campus backyards.

One such example is the University of California, Irvine, which is consistently rated among the greenest schools in America in *Sierra* magazine's "Cool Schools" rankings. And this is for good reason. Among UC Irvine's many accomplishments: the school has diverted 80 percent of its waste from landfills at the 56,000-student campus; it has two hundred electric car chargers and a fleet of twenty electric buses; it's phasing out disposable plastics; and its solar energy and energy efficiency programs have cut the power use in school buildings, labs, and cooling systems in half. Student-led initiatives have included a campus greenhouse, orchard, and vegetable gardens, as well as placing the school in the Bee Campus USA program, which ensures pollinator-friendly plants are used in landscaping.

Meanwhile, twelve US colleges and universities have become carbon neutral—the first of more than four hundred schools that have pledged to do so as part of Second Nature's Climate Leadership Network. The twelve are Colby College, Maine, carbon neutral in 2013 (first in the country); Middlebury College, Vermont, in 2016; Bates College, Maine, in 2016; Bowdoin College, Maine, in 2018 (thanks to the largest solar energy installation in the state); American University, Washington, DC, in 2018; Colgate University, New York, in 2019; University of San Francisco, California, in 2019 (hitting its goal thirty years ahead of schedule); Colorado College, Colorado, in 2020 (reducing emissions 75 percent through geothermal and solar power); Allegheny College, Pennsylvania, in 2020; Dickinson College, Pennsylvania, in 2020; Catawba College, North Carolina, in 2023; and Hampshire College, Massachusetts, in 2023 (my alma mater!).

Some schools find unusual ways to fight waste and provide breakthroughs others can emulate. Middle Tennessee State University ditched all its chemical cleaning products, most of which have toxic ingredients, for a radically different cleaner: Stabilized Aqueous Ozone—which is basically water with an extra oxygen atom. It's a bit like hydrogen peroxide, but without the toxicity or harshness. It disinfects and cleans all surfaces, then breaks down into regular water and oxygen. There are no emissions, and no residue that needs to be rinsed or cleaned off. A simple electrically powered wall unit generates the cleaner from plain water and air, so no disposable plastic bottles are needed, either. It's also safer to use, with no health risks and low carbon emissions. It has the same fresh, clean scent as the smell after a lightning storm, which is nature's way of creating aqueous ozone. Stanford University has adopted the same system, and others are following.

It's not just four-year colleges. Public primary and secondary schools are installing heat pumps in California, building solar arrays in the Midwest, and teaching wind turbine maintenance and solar energy installation in high schools and community colleges in wind country. At Fryberger Elementary School, near "Little Saigon" in conservative-leaning Orange County, the campus is 90 percent powered by solar arrays run by fourth graders, fed by kindergarten organic farmers, and has waste and recycling operations staffed by first graders. At this award-winning US Department of Education "Green Ribbon" school, the subjects of science, history, writing, reading, and civics are suffused with ecological themes and projects that guide kids and families to affordable, healthy, climate-friendly living.

Some innovative ways to tackle waste can be deceptively simple. For instance, students at the University of Minnesota Twin Cities campus volunteered to give up their weekends to open a free car-tire pressure clinic in their university town. Doesn't sound like much of a tactic in the war on waste, does it? But it is.

The students had decided to solve two problems. First, underinflated tires waste fuel, because the added friction of flat tires makes the engine work harder to cover the same distance. And second, most of us are lousy at keeping our tires properly inflated. The clinic would fix this for people—for free.

The townies loved the free clinic, and they also loved the math: the students calculated that every 130 cars that drove off with peak tire pressure saved gas and carbon emissions equal to replacing one fossil fuel–burning car with an EV. This also meant savings of a few hundred dollars a year at the gas pump for the average driver. And if you don't think little things like this add up when widely adopted, consider that properly inflated tires wear out more slowly and are also safer to drive—so more money and possibly lives are saved that way, too. And since tire wear is the process of tread rubbing off onto the road surface, peak tire pressure also cuts microplastic pollution, because these days, it's primarily the plastic, not the rubber, that meets the road. Those bits of tire are then swept away by wind and rain into drainage ditches, creeks, rivers, and oceans.

The effects of waste are far-reaching and often multiple in nature. So are the benefits of being less wasteful—even for something so seemingly minor as tire pressure: safety, health, money in our pockets, plus lower plastic pollution and climate-damaging emissions. Many Americans don't have the desire or the luxury to choose something solely because it is good for the environment, or the climate, or an endangered species, or some strangers facing hardship in a distant land. But if a choice also saves you money, or protects your family's health, or increases your quality of life, or just simply and plainly works better—or does all these things at once— well, that's a different story, isn't it? Whether you are a global warming hawk or climate change doubter, an avid hunter or a devoted vegan, a tree hugger or a lumberjack, a farmer or a city dweller (or both), making the less wasteful choices makes sense, and they add up.

This is the secret sauce that makes the Morris Model work.

"Schools and campuses have become sustainability leaders—in large part because our students demand it," says Troy Goodnough, sustainability director at the Morris campus. "It's a very exciting time. It's a hopeful time."

When Goodnough took the job in 2006, he was the first and only sustainability director for the University of Minnesota system, and one of a very few in the nation. It was a time when interest in conservation and sustainability were rekindling after a long period of dormancy following the last burst of interest in the 1960s and '70s, powered by the college students of that era. He jokes that his parents were never quite sure he had a real job.

In 2007, Morris was one of the early colleges to sign on to Second Nature's Climate Leadership Network, in which campuses across the country agreed to take a leadership role on the push to carbon neutrality, and then its more recent update, which adds a pledge to build community resilience in the face of a changing climate.

In addition to electricity from wind and solar, the Morris campus uses a solar thermal system to heat and cool multiple buildings and to heat its swimming pool. A biomass gasification reactor also turns wood, corncobs, and other agricultural waste into energy used for heating and cooling.

The school began expanding its efforts beyond the campus when students launched a composting operation for food waste in 2012, finding a way to do it simply in cold weather, when a number of experts had warned them it would be too difficult and expensive to try in Minnesota's frigid winters. The program was so successful that it expanded and began serving the surrounding community with drop-off composting. Food waste was being turned into valuable nutrients for local farms, diverting it from landfills, where it would have become a source of heat-trapping pollutants. After ten years, the students turned it over to the county, where former student Sydney Bauer, of Bert and Ernie name fame, is now running the new organics recycling program.

Meanwhile, Goodnough began looking for other ways campus and community could combine forces for sustainability, not just for local benefits, but also as a national model. "If we can do sustainability right in the upper Midwest, in Minnesota, with its extreme heat and its extreme cold, I think it could be figured out anywhere," he says.

The Morris Model—before it was a model for anything—didn't begin with talk of climate impact or sustainability, not at first, anyway. Goodnough wanted something easy and small, something that wouldn't be culturally divisive in a conservative rural area, but would instead bring people together across generations. The conversation began with energy efficiency, and how that could save the town money at a time when local government budgets were tight—which is, for the most part, all the time. Goodnough suggested to the Morris city leaders that they could work together on something that already had been successful on campus—converting to energy-efficient LED lighting. So the beginning was light bulbs: a tried-and-true method of saving volts and dollars, because the old incandescent bulb technology's main product is wasted heat, not light. There was pushback to phasing out the old-style bulbs in some parts of the country, where it became fodder for one of the least sensible fronts in the culture wars, but not in sensible, bottom-line-oriented Morris. The savings were just too crazy good: lighting bills go down by 75 to 80 percent when you make this switch. What does that mean in the real world? Well, one common 60-watt incandescent light bulb (sometimes called the "Edison bulb") costs about $58 to light up for a year, with average electric rates. Its LED equivalent costs $11.50. That's a $46.50 savings for just one bulb. You can see why switching them out at home makes sense—and why a city, with thousands of bulbs and much more powerful and expensive streetlighting to pay for, realized this was a jackpot. It was a big project for a city of five thousand, from Main Street outdoor lighting to every light fixture in city buildings. The university helped the city get grants to help fund the transition, and the local utility got on board. In the end, the conversion led to an $80,000 annual savings on the city utility bill, which, for a small budget, was a huge bit of found money.

And it put Morris ahead of the game. Ten years later, in 2023, incandescent bulbs were phased out in America for good (like they had been in most of the rest of the world long before). They are off store shelves now.

After that success, it was the city asking Goodnough what they could do next together. That led to solar rooftop installations on city hall, the city community center, the public library, and the liquor store, with four more installations planned. Next, the city, with the help of federal energy grants, put an electric bus into service, and hired Morris graduate Bauer to drive it, in addition to her compost duties and her wintertime gig running the popular (and very seasonal) Snowball Saloon, a gathering place built out of snow and ice, with holiday lights and a bonfire pit inside. The Morris city manager calls her the town's personal "force of nature." He also hired another Morris graduate, Griffin Peck, to become the city's first sustainability director. Both Morris graduates would play important roles in developing Morris's ambitious strategic plan for a renewable future.

But first there was one more profound influence on the process—a sister-city program begun by the university with a rural farm town in Germany's Westphalia region called Saerbeck. The German municipality had launched its own sustainability and energy independence program in 2009, and had built so much renewable energy capacity—solar, wind, biogas—that it made four times the electricity it needed. Because the program was established as a cooperative, and each citizen of Saerbeck owned a share, everyone began receiving checks for all the excess electricity the town was able to sell. Goodnough convinced city manager Hill to join a group from the university visiting Saerbeck, and the Morris leader ended up hitting it off with the mayor there, a town elder who said that what they had done for renewable energy and conservation accomplished two goals: it made economic sense by saving and then making money, and it built something for his grandkids—a better future. Hill looked at what a town very much like Morris had accomplished in Germany, and it changed him in a way that a bunch of upstart college students never could have, at least not on their own. Goodnough said it was a significant turning point for the Morris Model, with Hill saying, "We can do this, too."

This led to the adoption of the strategic plan for the city's sustainability future:

1. Produce 80 percent of the energy consumed in the county by 2030.
2. Reduce energy consumption by 30 percent by 2030.
3. End landfilling of waste generated within the county by 2025.

This Morris Model formula works because it focuses on the economic benefits of cutting waste and embracing efficiency independent of their value in fighting climate disruption. Each one of those broad goals comes with a payoff—financially, in building resilience, and in turning trash into treasure. It's not exactly hidden from view—they talk about it all the time, because it's a direct result of becoming energy efficient and renewable-based—but lowering the climate footprint of the town is not the animating principle for the farmers for any of this. They like renewable energy because it saves the town money and because it's a source of pride and independence.

Goodnough calls Bauer, Peck, and their peers at Morris sustainability rock stars, part of a broad base of student activism at the school. Their efforts have been critical in pushing the campus toward ever more ambitious goals, and their activism continues after graduation as they become leaders in working to save the planet, or at least striving to avert as much disaster as possible. He has contact with his sustainability counterparts at schools around the country and the world, and he hears the same thing all the time: this current generation of students and recent graduates are committed to change, and may be our greatest asset in turning the tide on the environment, climate, and waste.

There is plenty of evidence in support of this claim. Look no further than the "Montana 16," a group of young people, ages two to eighteen, who sued their state for failing to act against global climate change, while ignoring a state constitution that explicitly guarantees the right to a "clean and healthful environment." They filed the suit in 2020, something children have been doing in other states as well, asking courts to compel the government to take meaningful action against the emissions driving climate disruption. The young plaintiffs argue they have the most to lose—it's their future that might end up in ashes—and therefore they should be granted standing in court. In 2023, despite every legal trick in the book deployed against it, the Montana case became the first to go to trial in the country. The kids won, a watershed moment that could have far-reaching effects and that, whatever happens, has energized youth activists worldwide.

"We are heard!" Kian Tanner, one of the Montana 16, said after the verdict. "Frankly, the elation and joy in my heart is overwhelming in the best way. We set the precedent not only for the United States, but the world."

Next consider Nalleli Cobo, who became an activist while still in elementary school, attending meetings and rallies with her mom because of concerns over pollution and waste from an oil well across the street from their home. She lives in southern Los Angeles, a city that, among all its other more famous attributes, is also the largest working urban oil field in the country, with creaking, reeking, bobbing oil wells tucked into neighborhoods all over the metropolitan area.

Cobo, the daughter of Colombian and Mexican immigrants, had been suffering from headaches, nosebleeds, stomach pain, asthma, and heart palpitations, which the family felt certain were caused by the noxious fumes emanating from the well across the street. She gave her first public speech at age nine.

And Cobo was a natural. She spoke of the terrible fumes she grew up with. How her parents had to close the windows, even on the hottest days, to avoid the dizzying smells. Her poise and moving speech captivated and motivated the adults in the neighborhood to join her in reporting the constant pollution to the city council. Then she went door to door, drumming up support. At fifteen she cofounded and became spokesperson for a grassroots environmental group, People Not Pozos (oil wells), and another organization, the South Central Youth Leadership Coalition, which focused on environmental racism in LA, calling out oil companies for running polluting wells for decades in primarily Black, brown, and low-income neighborhoods. She filed complaints with the state pollution control agency, and her appeals persuaded Physicians for Social Responsibility Los Angeles to arrange for a toxicologist to test the air quality around the well. It proved to be toxic.

Cobo gave media interviews, met with Al Gore and Bernie Sanders, and sued the city for environmental racism for disproportionately approving oil drilling permits in Black and Latino communities. She won the case—forcing the city and all of Los Angeles County to reform policies on oil extraction in neighborhoods, including a ban in the county on new wells. The well across the street from Cobo's house was shut down and the owner's corporate executives were charged with twenty-four environmental crimes.

Nalleli Cobo is a remarkable person, but her superpower was one we all have: she showed up. She went to her neighbors. She went to community meetings. She went to the city council and the county board of supervisors. She complained and pleaded and inspired (and sometimes shamed) the powerful into taking action and doing their jobs. This is part of the path from waste and pollution and environmental injustice everyone can take, everyone *must* take, if we want to get out of this mess. It's what Laura Marston in Maine learned she could do, testifying as a small-business owner about plastic waste. It's what Jamiah Hargins did when he went to the state legislator for his part of LA and explained his vision for urban microfarms—and ended up with a grant to make it happen. The simple act of asking for things that make a difference works. It can be asking the school board for bike racks so kids can pedal to school instead of being driven. Or you could encourage your town council to plant shade trees on barren streets where the summer sun is relentless. You could ask your favorite restaurant to stop giving out plastic and foam disposables—and nicely tell them you'll have to eat elsewhere if they can't oblige. Little things add up, and they can happen if we just ask. And push. And show up, like Nalleli Cobo.

In the midst of her victories, the years of toxic exposure exacted their toll: at age nineteen, Cobo was diagnosed with reproductive cancer, and though chemotherapy, radiation, and multiple surgeries eventually left her cancer-free, she was also left unable to bear children. She continues her work and her activism while attending college, and in 2022 she received the Goldman Environmental Prize for her grassroots leadership for environmental justice. That same year, singer-songwriter Billie Eilish invited eight young climate justice activists to join her for the cover story that *Vogue* magazine wanted to do on her, and Cobo was one of the eight.

"We have the right to breathe clean air, to open our windows in our own homes," she said during the magazine photo shoot. "I think everyone has a very important role to play when it comes to the climate crisis. I think of it like we're all a symphony. Whether you're playing a small instrument like the flute or a big one like the bassoon, you're instrumental, you're critical to this movement, and you do hear it when an instrument falls out."

A Waste Less Future

There is a path to a future that's not total garbage. The city of Morris and its university partner are on it together. So is Peachtree City. And Maine, and Sarah Nichols, Jamiah Hargins, Jenna Jambeck, and Nalleli Cobo. We all can be.

Of course, there's no fairy-tale ending. Putting a stop to our wasteful ways will not immediately undo the damage our waste has already done—the plastic ocean, the toxic chemicals, the climate-disrupting pollutants. That will take generations, and those in the future will not thank us for taking so long to act. But what is happening in Morris, and on campuses and in communities nationwide, can stop these crises of waste from getting worse in the here and now, and in the future for our children and grandchildren. That is the challenge now for the generations living today.

I wrote at the beginning of this book that the time for hope was past, that it was time to get off our asses and get it done now or else. I stand by that, but after meeting the people depicted in this book, after seeing what they can do and how they inspire others, and knowing that there are so many others out there taking on these challenges, too, I have to say it: I am hopeful. More than I have ever been. Taking action is, in fact, an act of hope. We can do this. We can fix the waste driving all these crises. We can. As Al Gore said just a few months before I sat down to write these lines, "The will to act is itself a renewable resource."

Most people are on board for this. The Pew poll of climate attitudes tells us so. Ninety percent of Greek people surveyed said climate change is a major threat. In South Korea, 86 percent agreed. In France, it was 83 percent, Japan 75 percent, the Philippines 67 percent, Australia 60 percent. And here in the US? Fifty-nine percent say climate change is a major threat, and only 16 percent say it's not a threat at all. That's a clear majority who get it. A landslide, really.

"I'm hopeful," Troy Goodnough says. "Look at what we've done here, in a place no one expected to see sustainability and renewable energy and zero waste flourish. There's no way you can look at this and not think anything is possible."

———

The *Total Garbage* story doesn't end here—it's a work in progress, as is the search for solutions to our waste and the big crises it drives. The innovators and leaders depicted in this book are not alone—there are so many others finding, building, and pursuing solutions.

So let's continue weaving this story together. I invite you to send me your solutions on waste, plastic pollution, energy, transportation, and climate, or great ones you've heard about in your community, state, school, or local businesses. Let's find the next chapter of *Total Garbage* together!

To access the *Total Garbage* Reading Guide and further resources, visit edwardhumes.com/total-garbage-readers-guide.

What You Can Do Right Now about Waste: The Master List

PLASTICS WASTE

- Don't accept plastic bags when you shop anywhere.
- Bring your own bags and containers to the market and everywhere you shop.
- Just say no to plastic utensils, plastic takeout containers, and plastic-foam anything.
- Bring your own containers to restaurants to use instead of doggie bags (or foam takeout containers).
- Avoid all single-use plastic beverage bottles.
- Buy alternative forms of common products to avoid plastic containers: toothpaste tablets, bamboo toothbrushes, shampoo and conditioner bars, laundry detergent bars or sheets. Or buy these liquid products in bulk and bring your own bottles to a refill store.
- Avoid plastic wrap, plastic bags, single-use plastic containers, and conventional wax paper for food storage. Use reusable glass or stainless-steel containers instead. Saving old jars is a low-cost option.
- Use reusable cups for coffee or any other single-serve beverage you buy.

- Keep a zero-waste kit in your car, bike bag, backpack, or whatever you use when you go out so you are ready with reusable alternatives to common disposables.
- Recycle the packaging you do buy.
- If you live in one of the ten container deposit states, bring those items to a redemption center, where they are far more likely to be recycled. The ten are: California, Connecticut, Hawaii, Iowa, Maine, Massachusetts, Michigan, New York, Oregon, and Vermont.
- Volunteer for beach, park, and river cleanups.
- Use the Debris Tracker app to help solve the plastic pollution crisis.
- Form your own neighborhood Owen's List to get hard-to-recycle items to places that will recycle them.
- Support proposals in your state and community to ban disposable plastics, to adopt extended producer responsibility laws, and to adopt container-deposit laws. Also vote for politicians who favor such policies.
- If your community or local school district makes little or no effort to recycle, reuse, or reduce, lobby your elected officials to enter the twenty-first century and start dealing with the plastic waste problem. Bring it up at the next council or board meeting and get the ball rolling!

ENERGY WASTE

GAS STOVE EMISSIONS

- Cook with the windows open.
- Use your range hood and cook on the back burners, where more fumes are captured by the hood.
- Use a window fan set on exhaust while cooking.
- Close bedroom doors and other spaces while cooking to keep stove toxins out of those rooms.

- Use a countertop induction burner and other electric appliances such as toaster ovens as your main cooking tools, with the gas stove as your backup or when you need multiple burners.
- Replace your gas stove with an electric stove, conventional or induction. Such a purchase could be eligible for a rebate under the Inflation Reduction Act of 2022.

WHAT YOU CAN DO TO "SQUEEZE THE JUICE"

Big Projects

- Install rooftop solar panels.
- Upgrade home heating and cooling systems to an energy- and money-saving heat pump.
- Convert a gas or conventional electric hot water heater to a heat-pump model.
- Upgrade home insulation, air-sealing, or windows to increase energy efficiency.
- The federal Inflation Reduction Act offers rebates of up to $14,000 for a long list of home energy upgrades.

Quick and Easy Energy Savings

- Use efficient LED light bulbs.
- Power down computers at night instead of letting them sleep.
- Use ceiling fans or window fans to lower the need for air-conditioning.
- Cool and heat only rooms in use rather than the entire home.
- Turn the thermostat down to the low sixties at bedtime in winter.
- Use shade trees, blinds, or awnings to passively cool the home and reduce the load on air-conditioning in the summer.
- Unplug "vampire electronics" when not in use: cable boxes, game consoles, TVs, streaming boxes, phone chargers. Putting several

such devices on a single power strip with a switch makes unplugging easy.

- Set your washing machine to cold water, wash full loads, and consider line-drying clothes on nice days.
- Cut your time in the shower.
- Replace grass with less-water-consuming plants.
- Use the dishwasher instead of handwashing.

TRANSPORTATION

- Walk or bike for short trips instead of driving.
- Find ways to electrify your non-human-powered trips. Consider using an e-bike or NEV (electric golf cart) for trips under 10 miles or so instead of burning gasoline to get there.

AVOID FOOD WASTE

- Make a meal plan. "Shop" your refrigerator and pantry and make a plan to use what you have, and then make a list of the ingredients you need so you avoid buying unnecessary items.
- When buying fresh foods, buy only what you need for a few days.
- Use recipes as guidelines, not strict rules, so your fridge and pantry don't pile up with ingredients you don't often need or use.
- Serve modest portions to avoid plate waste. Make veggies the largest portion and animal proteins the smallest. You can always go back for seconds.
- Repurpose leftovers.
- Repurpose food "waste." The stems, leaves, and peels of many veggies—beets, cauliflower, onion skins, carrot ends and peelings, and celery tops, to name a few—can be made into vegetable stock for soup making.

- Make soup your go-to main dish on work nights. A big pot of soup can provide multiple veggie-forward meals for the week with very little wasted. All the nutrition stays in the pot.
- Buy fresh and unpackaged food at your local farmers market.
- Plant your own vegetable garden or microfarm. Whether it's a big chunk of your back or front yard, a patch at the local community garden, or some potted veggies or a mini-greenhouse on your apartment porch or balcony, growing your own food cuts waste, saves money, and delivers better flavor and nutrition—and it's fun! The energy and water savings are particularly dramatic if you are replacing grass lawns with vegetables or native plants.

SHOPPING WITHOUT WASTE

- Shop at your local no-waste store for bulk purchases of common items like laundry detergent and shampoo using your own containers. Find a zero-waste or refill store near you at litterless.com.
- Reuse, reuse, reuse! Beyond your reusable tote bags, focus on ditching disposables. That means: reusable water bottles and coffee cups. Paperless paper towels. Reusable razors. Almost every disposable has a reusable alternative. Even coffee filters!
- Package-free is the best, but some packages are better than others. If you must buy something in disposable containers, look for packages of cardboard, metal, or glass—all much more recyclable or reusable than plastic.
- Join the secondhand and borrowing economies. Borrowing, trading, and buying secondhand at thrift stores, Buy Nothing groups, Facebook Marketplace, and Craigslist are all great waste cutters and money savers. Buying clothing this way is particularly powerful.
- Buy used or refurbished electronics instead of new.
- Repair, rather than replace, when possible.

Author's Note and Acknowledgments

The seeds of *Total Garbage* were planted years ago at a sustainability summit convened by Walmart. One speaker had a profound effect on me. They argued that the key to advancing the business case for sustainability was for companies to focus on waste. The idea was that even investors and corporate boards averse to anything environmentalists wanted would cheer on waste reduction as a source for cost savings and increased profits. Yet the results would be just as green: less energy waste and more efficiency; less burning of fossils fuels; less plastic packaging; more recycling and composting; and less landfill waste. To be sure, the goal wasn't to save the world, but simply to call out the financial payoffs for some modest steps to make a big company less unsustainable. But I was floored.

I sensed a very big truth tucked inside this modest proposal: Going after waste, not just in retail but throughout the human-made world, could unify and reframe debate and action on *all* our big environmental crises. Waste could be the Rosetta Stone of conservation, the universal language for united acting on climate, pollution, renewables, green transportation, producer responsibility—everything. My first foray into exploring this story focused on the hidden world of our garbage in *Garbology: Our Dirty Love Affair with Trash*. Then I wanted to tell a story of solutions: all the ways we waste ourselves into oblivion, and how we can un-waste ourselves to a brighter future. My goal was to tell the stories of people and communities doing the hard, inspiring, joyful, heroic work of making sure the title of this book stays more metaphor than reality.

I hope it will always be so.

Total Garbage could not have been written without the inspiring people depicted on its pages, and their generosity in sharing their knowledge, insights, and patience with my many questions. I particularly wish to thank Ryan Metzger, Dr. Jenna Jambeck, Sarah Nichols, Chef Chris Galarza, Chef Rachelle Boucher, Brady Seals, Amory Lovins, Jamiah Hargins, Janet and Jeff Crouch, Nancy Lawson, Zero-Waste Chef Anne-Marie Bonneau, Dr. Rachel Schattman, Ashley Walsh, Chris Bradshaw, Mayor Kim Learnard, Laura Marston, Troy Goodnough, Sydney Bauer, Blaine Hill, and Griffin Peck.

I also want to express deep gratitude to my very patient and skillful editor, Lauren Appleton; my agent and dear friend, Susan Ginsburg at Writers House; and the superpower-endowed Catherine Bradshaw, also of Writers House. Thank you all for helping me make sure *Total Garbage* is not total garbage!

Finally, there would be no *Total Garbage* without the total love, support, and occasional mockery from my children, Eben and Gabrielle, and my partner in all things, Donna Wares. "Thank you" will never be enough. May we always *Roooooooooooooo!* together with the greyhounds.

Notes

PROLOGUE: THE CREDIT CARD

xiii **You swallowed 285 pieces of plastic today:** Dalberg Advisors, Wijnand de Wit, and Nathan Bigaud, *No Plastic in Nature: Assessing Plastic Ingestion from Nature to People*, an analysis for World Wildlife Fund by Dalberg and the University of Newcastle, Australia, 2019.

xv **at least 2,400 known toxins:** Valerie Denney, *An Introduction to Plastics and Toxic Chemicals: How Plastics Harm Human Health and Poison the Circular Economy*, International Pollutants Elimination Network, November 2022, https://ipen.org/documents/introduction -plastics-and-toxic-chemicals; Helene Wiesinger, Zhanyun Wang, and Stefanie Hellweg, "Deep Dive into Plastic Monomers, Additives, and Processing Aids," *Environmental Science and Technology* 55 (June 7, 2021): 9339–51, https://doi.org/10.1021/acs .est.1c00976.

xv **PFAS, also known as "forever chemicals":** PFAS is short for "per- and polyfluoroalkyl substances." These "forever chemicals" are described this way by the US Environmental Protection Agency:
  • PFAS are widely used, long lasting chemicals, components of which break down very slowly over time.

- Because of their widespread use and their persistence in the environment, many PFAS are found in the blood of people and animals all over the world and are present at low levels in a variety of food products and in the environment.
- PFAS are found in water, air, fish, and soil at locations across the nation and the globe.
- Scientific studies have shown that exposure to some PFAS in the environment may be linked to harmful health effects in humans and animals.
- There are thousands of PFAS chemicals, and they are found in many different consumer, commercial, and industrial products. This makes it challenging to study and assess the potential human health and environmental risks.

xv **cancers of the digestive system:** Tomotaka Ugai et al., "Is Early-Onset Cancer an Emerging Global Epidemic? Current Evidence and Future Implications," *Nature Reviews Clinical Oncology* 19 (October 2022): 656–73, https://doi.org/10.1038/s41571 -022-00672-8. Findings are summarized in Brigham and Women's Hospital Communications, "Dramatic Rise in Cancer in People under 50," *Harvard Gazette*, September 8, 2022, https://news.harvard .edu/gazette/story/2022/09/researchers-report-dramatic-rise-in-early -onset-cancers/.

xv **400 million tons a year and counting:** "Our Planet Is Choking on Plastic," United Nations Environment Programme, 2023, https:// www.unep.org/interactives/beat-plastic-pollution/.

xxii **Ayana Elizabeth Johnson put it this way:** "2023 Middlebury Commencement Address by Dr. Ayana Elizabeth Johnson," May 28, 2003, Middlebury, VT, https://www.middlebury.edu /announcements/2023/06/2023-middlebury-commencement-address -dr-ayana-elizabeth-johnson.

CHAPTER 1: OUR DISPOSABLE AGE

6 **inventor Dr. Leo Hendrik Baekeland:** "Science: At Ithaca," *Time*, September 22, 1924, https://content.time.com/time/magazine/article /0,9171,719175-1,00.html.

CHAPTER 2: TRASH GENIUS

21 **the world's only official Trash Genius:** "Jenna Jambeck: Environmental Engineer, Class of 2022," award page, MacArthur Foundation, https://www.macfound.org/fellows/class-of-2022 /jenna-jambeck.

22 **nearly 9 million tons:** Jenna R. Jambeck et al., "Plastic Waste Inputs from Land into the Ocean," *Science* 347, no. 6223 (February 13, 2015): 768–71, https://doi.org/10.1126/science.126035; Roland Geyer, Jenna R. Jambeck, and Kara Lavender Law, "Production, Use, and Fate of All Plastics Ever Made," *Science Advances* 3, no. 7 (July 19, 2017), https://doi.org/10.1126/sciadv.1700782.

29 **National Sword did not cause a crisis:** Edward Humes, "The US Recycling System Is Garbage: China Doesn't Want Our Crap Anymore, and Who Can Blame Them," *Sierra*, June 26, 2019, https://www.sierraclub.org/sierra/2019-4-july-august /feature/us-recycling-system-garbage; Edward Humes, "Zeroing Out Zero Waste: A Conversation with David Allaway, Recycling Heretic," *Sierra*, June 26, 2019, https://www .sierraclub.org/sierra/2019-4-july-august/feature/zeroing-out-zero -waste.

30 **recycle only about half of all aluminum beverage cans:** "Aluminum: Material-Specific Data," US Environmental Protection Agency, last updated November 22, 2023, https://www.epa.gov/facts-and -figures-about-materials-waste-and-recycling/aluminum-material -specific-data.

**1960–2018 Data on Aluminum Metals in MSW by
Weight (in Thousands of US Tons)**

Management Pathway	1960	1970	1980	1990	2000	2005	2010	2015	2017	2018
Generation	340	800	1,730	2,810	3,190	3,330	3,510	3,670	3,820	3,890
Recycled	-	10	310	1,010	860	690	680	670	600	670
Composted	-	-	-	-	-	-	-	-	-	-
Combustion with Energy Recovery	-	-	30	300	390	410	440	510	550	560
Landfilled	340	790	1,390	1,500	1,940	2,230	2,390	2,490	2,670	2,660

31 the Great Pacific Garbage Patch: From "What Is the Great
Pacific Garbage Patch?," National Oceanic and Atmospheric
Administration, https://oceanservice.noaa.gov/facts/garbagepatch
.html:

> While "Great Pacific Garbage Patch" is a term often used by the
> media, it does not paint an accurate picture of the marine debris
> problem in the North Pacific ocean. Marine debris concentrates in
> various regions of the North Pacific, not just in one area. The exact size,
> content, and location of the "garbage patches" are difficult to accurately
> predict.
>
> The name "Pacific Garbage Patch" has led many to believe
> that this area is a large and continuous patch of easily visible
> marine debris items such as bottles and other litter—akin to a literal
> island of trash that should be visible with satellite or aerial photographs.
> This is not the case. While higher concentrations of litter items can be
> found in this area, much of the debris is actually small pieces of floating
> plastic that are not immediately evident to the naked eye.
>
> Ocean debris is continuously mixed by wind and wave action and
> widely dispersed both over huge surface areas and throughout the top
> portion of the water column. It is possible to sail through "garbage

patch" areas in the Pacific and see very little or no debris on the water's surface. It is also difficult to estimate the size of these "patches," because the borders and content constantly change with ocean currents and winds.

CHAPTER 3: MESSAGE IN A BOTTLE

42 **A law is nothing without implementation:** "Extended Producer Responsibility Program for Packaging," Maine Department of Environmental Protection, details and timeline for implementation, https://www.maine.gov/dep/waste/recycle/epr.html.

43 **a new career—bottle detective:** Jane Busch, "Second Time Around: A Look at Bottle Reuse," *Historical Archaeology* 21, no. 1 (1987): 67–80, http://www.jstor.org/stable/25615613.

45 **Candler sold the bottling rights in 1899:** *125 Years of Sharing Happiness: A Short History of the Coca-Cola Company*, Coca-Cola, https://www.coca-colacompany.com/content/dam/journey/us /en/our-company/history/coca-cola-a-short-history-125-years -booklet.pdf.

48 **mandatory deposits and outright bans on nonreturnable glass bottles:** Bartow J. Elmore, "The American Beverage Industry and the Development of Curbside Recycling Programs, 1950–2000," *Business History Review* 86, no. 3 (Autumn 2012): 477–501, https:// doi.org/doi:10.1017/S0007680512000785; Andrew Boardman Jaeger, "Forging Hegemony: How Recycling Became a Popular but Inadequate Response to Accumulating Waste," *Social Problems* 65, no. 3 (August 2018): 395–415, https://doi.org/10.1093/socpro/spx001.

CHAPTER 4: RING OF FIRE

80 **The Facts about Harmful Gas Stove Emissions:** Brady Anne Seals and Andee Krasner, *Health Effects from Gas Stove Pollution*, Rocky Mountain Institute, Physicians for Social Responsibility, Mothers

Out Front, and the Sierra Club, 2020; Taylor Gruenwald et al., "Population Attributable Fraction of Gas Stoves and Childhood Asthma in the United States," *International Journal of Environmental Research and Public Health* 20, no. 1 (2023): 75, https://doi.org/10.3390/ijerph20010075; Yannai S. Kashtan et al., "Gas and Propane Combustion from Stoves Emits Benzene and Increases Indoor Air Pollution," *Environmental Science and Technology* 57, no. 26 (June 15, 2023): 9652–63, https://doi.org/10.1021/acs.est.2c09289; Eric D. Lebel et al., "Methane and NO_x Emissions from Natural Gas Stoves, Cooktops, and Ovens in Residential Homes," *Environmental Science and Technology* 56, no. 4 (January 27, 2022), https://doi.org/10.1021/acs.est.1c04707.

CHAPTER 5: TAKING THE HEAT

90 **the gas industry has known:** Rebecca John, "Industry Knew about Gas Stoves' Air Pollution Problems in Early 1970s," DeSmog, March 2, 2023, https://www.desmog.com/2023/03/02/american-gas-association-knew-stoves-air-pollution-1970s/.

90 **1981 finding by the EPA:** In 1981 an EPA committee on indoor pollutants found "an association between gas cooking and the impairment of lung function in children." National Research Council (US) Committee on Indoor Pollutants, *Indoor Pollutants* (Washington, DC: National Academies Press, 1981), https://www.ncbi.nlm.nih.gov/books/NBK234058/.

95 **That blue light doesn't cook anything:** The supposed cooking power of a gas stove's blue flames is a myth and misunderstanding about the physics of cooking. That blue is just visible light. Any sufficiently hot substance can give off light, whether the metal filament inside an electric light bulb or a fuel on fire. But the heat causes that light, not the other way around. Heat from burning gas

on a stove comes from something different: infrared radiation (IR), sometimes called infrared light, even though it's invisible. On a stovetop, the IR causes a stream of heated air and combustion gases to rise up from the burner, where it heats the pan above. On a grill, the hot air, not the flames, cooks the food directly, which is why electric barbecues and broilers work just as well as their gas competitors. IR from burning gas also directly strikes the pan or the food on the grill, but the effect is weaker than the hot air because IR spreads in all directions, while the hot air just rises straight up. This is why you can be warmed by a campfire by being next to it, but you can only cook on top of it.

Although they are not familiar to most consumers, gas infrared burners are available. The technology was developed in the early 1980s, when it seemed pollution regulations for gas stoves were imminent and the industry sought a greener solution that didn't completely eliminate gas stoves. Infrared burners dispense with the conventional stove's familiar metal ring, from which the blue flames spring, and instead use a ceramic plate with many more and smaller perforations. As the gas burns, the plate heats up and glows red-orange, focusing the IR and hot gases more directly into the pot above, while bleeding off less heat outward into the room. That makes gas IR burners significantly less wasteful than conventional ones, using 40 to 50 percent less gas to perform the same cooking, and cutting pollutants by similar percentages. IR burners remain a niche product, though, for several reasons: the regulations never came, and the ceramic plates are much more expensive than the old-school $6 metal burner rings that come even with $3,000 stoves. Perhaps the hardest pill for the marketers of gas stoves to swallow, though, is that with IR gas ranges, there are no blue flames, or any very visible flames at all. The red glow looks and mostly acts like . . .

conventional electric ceramic stoves. Which, unsurprisingly, are also known as electric infrared stoves.

With this new invention, the gas industry itself proved the celebrated blue flames were a meaningless light show. *Everything* cooks with infrared—also known as thermal energy—the only difference being how the heat is created and where. The reason induction stoves are so efficient is that the infrared begins in the pan itself, minimizing wasted energy and eliminating any part that creates waste pollutants.

CHAPTER 6: SQUEEZING THE JUICE

102 **family home and a giant science experiment:** "Amory's Private Residence," Rocky Mountain Institute, https://rmi.org/about/office-locations/amory-private-residence/; Ben Adler, "Amory Lovins' High-Tech Home Skimps on Energy but Not on Comfort," *Grist*, July 25, 2014, https://grist.org/climate-energy/amory-lovins-high-tech-home-skimps-on-energy-but-not-on-comfort/.

106 **a movement called Passive House:** Somini Sengupta, "Building Better Buildings," *New York Times*, April 14, 2023, https://www.nytimes.com/2023/04/14/climate/passive-house-climate.html.

107 **up to 90 percent less energy than a conventional home:** "What Is a Passive House?," Passipedia: The Passive House Resource, last modified September 25, 2022, https://passipedia.org/basics/what_is_a_passive_house.

108 **bought a new home built to the Passive House standard:** Keith Shortall, "This Maine Home Can Stay 70 Degrees without a Furnace, Even When It's Freezing Outside," Maine Public, January 25, 2023, https://www.mainepublic.org/environment-and-outdoors/2023-01-25/this-maine-home-can-stay-70-degrees-without-a-furnace-even-when-its-freezing-outside.

111 **We waste two-thirds of the energy we consume:** Ann Parker, "Charting the Nation's Energy Use," Lawrence Livermore National Laboratory, February 2021, https://str.llnl.gov/2021-02/simon.

CHAPTER 7: CHUTES AND LADDERS

115 **By 2023, BlocPower had raised more than $250 million:** "BlocPower Announces $150 Million Financing, Is Honored by Vice President Harris, Unveils Corporate Rebrand," BlocPower, press release, March 1, 2023, https://www.blocpower.io/posts/series-b -financing-rebrand-vp-harris.

115 **reports of more meager savings or outright energy cost increases in some:** Lee Harris, "Energy Insufficiency," *American Prospect*, September 6, 2023, https://prospect.org/environment /2023-09-06-energy-insufficiency-blocpower/.

117 **known as a Sankey diagram:** "Energy Flow Charts: Charting the Complex Relationships among Energy, Water, and Carbon," Lawrence Livermore National Laboratory, https://flowcharts.llnl gov; "How to Read an LLNI Energy Flow Chart (Sankey Diagram)," Lawrence Livermore National Laboratory, April 19, 2016, YouTube video, 3:02, https://www.youtube.com/watch?v=G6dIvECRfcI.

122 **the economic case for renewables isn't even close:** Angel Adegbesan, "Solar Is Now 33% Cheaper Than Gas Power in US, Guggenheim Says," Bloomberg, October 3, 2022, https://www.bloomberg.com /news/articles/2022-10-03/solar-is-now-33-cheaper-than-gas-power-in -us-guggenheim-says; Max Roser, "Why Did Renewables Become So Cheap So Fast?," Our World in Data, December 1, 2020, https:// ourworldindata.org/cheap-renewables-growth.

123 **2,000 gigawatts of solar, wind, and energy storage proposed:** "Grid Connection Requests Grow by 40% in 2022 as Clean Energy Surges, Despite Backlogs and Uncertainty," Berkeley Lab, April 6, 2023, https://emp.lbl.gov/news/grid-connection-requests-grow-40-2022-clean.

124 **the more solar and wind farms we build:** John J. Berger, "Mark Jacobson: How One American Atmospheric and Climate Scientist Created Clean Energy Roadmaps for 50 U.S. States—and 139 Nations," *Sustain Europe*, Spring/Summer 2019, https://web.stanford .edu/group/efmh/jacobson/Articles/I/19-04-SustainEurope.pdf; Stacy Clark, "A Decidedly Impartial Review of Mark Jacobson's 100% Clean, Renewable Energy and Storage for Everything," *Renewable Energy World*, April 22, 2021, https://www.renewableenergyworld .com/wind-power/a-decidedly-impartial-review-of-mark-jacobsons -100-clean-renewable-energy-and-storage-for-everything.

125 **Nationwide, community solar gardens generate 3,200 megawatts:** "Community Solar," National Renewable Energy Laboratory, December 2021, https://www.nrel.gov/state-local-tribal/community -solar.html.

128 **Major power blackouts are more frequent:** "Surging Weather- Related Power Outages," Climate Matters, September 14, 2022, https://www.climatecentral.org/climate-matters/surging-weather -related-power-outages.

130 **"vampire electronics":** Pierre Delforge, Lisa Schmidt, and Steve Schmidt, *Home Idle Load: Devices Wasting Huge Amounts of Electricity When Not in Active Use*, Natural Resources Defense Council, May 2015, https://www.nrdc.org/sites/default/files/home -idle-load-IP.pdf; Brian Palmer, "Energy Vampires: Keep Your Devices from Wasting Energy and Money," Natural Resources Defense Council, May 2, 2022, https://www.nrdc.org/stories /energy-vampires-keep-your-devices-wasting-energy-and-money.

CHAPTER 8: STICK A FORK IN IT

134 **Crop Swap LA, would start killing lawns:** Crop Swap LA (website), https://www.cropswapla.org.

145 **38 percent was either unsold or uneaten:** "In the U.S., 38% of All Food Goes Unsold or Uneaten—and Most of That Goes to Waste," ReFED, food-waste analysis, https://refed.org/food-waste /the-problem/.

152 **Industrial farming has sapped the nutrition:** Cheryl Long, "Industrially Farmed Foods Have Lower Nutritional Content," *Mother Earth News*, May 6, 2009, https://www.motherearthnews .com/sustainable-living/nature-and-environment/nutritional-content -zmaz09jjzraw/; Stacey Colino, "Fruits and Vegetable Are Less Nutritious Than They Used to Be," *National Geographic*, May 3, 2022, https://www.nationalgeographic.co.uk/environment-and -conservation/2022/05/fruits-and-vegetables-are-less-nutritious-than -they-used-to-be; Donald R. Davis, Melvin D. Epp, and Hugh D. Riordan, "Changes in USDA Food Composition Data for 43 Garden Crops, 1950 to 1999," *Journal of the American College of Nutrition* 23, no. 6 (December 23, 2004): 669–82, https://doi.org /10.1080/07315724.2004.10719409; Erica Eberl et al., "Temporal Change in Iron Content of Vegetables and Legumes in Australia: A Scoping Review," *Foods* 11, no. 1 (December 27, 2021), https://doi .org/10.3390%2Ffoods11010056.

159 **"externalities" at *$2 trillion*:** Rockefeller Foundation, *True Cost of Food: Measuring What Matters to Transform the U.S. Food System*, July 2021, https://www.rockefellerfoundation.org/wp-content /uploads/2021/07/True-Cost-of-Food-Full-Report-Final.pdf.

CHAPTER 9: THE CAR OF THE FUTURE ISN'T WHAT YOU THINK

164 **half of our vehicle trips:** Vehicle Technologies Office, "More Than Half of All Daily Trips Were Less Than Three Miles in 2021," Office of Energy Efficiency and Renewable Energy, March 21, 2022,

https://www.energy.gov/eere/vehicles/articles/fotw-1230-march-21
-2022-more-half-all-daily-trips-were-less-three-miles-2021.

168 **fifty-three thousand premature deaths:** Fabio Caiazzo et al., "Air
Pollution and Early Deaths in the United States. Part I: Quantifying
the Impact of Major Sectors in 2005," *Atmospheric Environment* 79
(November 2013): 198–208, https://doi.org/10.1016/j.atmosenv
.2013.05.081.

168 **the economics of cars:** AAA, *Your Driving Costs*, 2022,
https://newsroom.aaa.com/wp-content/uploads/2022/08/2022
-YourDrivingCosts-FactSheet-7-1.pdf.

169 **Fatal car crashes in the United States:** "Preliminary Semiannual
Estimates," National Safety Council Injury Facts, https://injuryfacts
.nsc.org/motor-vehicle/overview/preliminary-estimates/; "Fatality
Facts 2021: Yearly Snapshot," Insurance Institute of Highway Safety,
May 2023, https://www.iihs.org/topics/fatality-statistics/detail
/yearly-snapshot.

169 **emergency room every *ten seconds*:** Based upon the CDC count of
3.4 million ER visits from car crashes yearly, which works out to
about one every nine to ten seconds. Danielle Davis and Christopher
Cairns, "Emergency Department Visit Rates for Motor Vehicle
Crashes by Selected Characteristics: United States 2017–2018,"
NCHS Data Brief no. 410, June 2021, https://www.cdc.gov/nchs
/products/databriefs/db410.htm. Also of interest: The National
Safety Council counts "medically consulted injuries" from car
crashes, which may be a paramedic at the scene, an ER visit, or a
checkup after the crash. There are 5.4 million medically consulted
car crash injuries a year, which would work out to one every six
seconds. See "Overview: Introduction," National Safety Council
Injury Facts, https://injuryfacts.nsc.org/motor-vehicle/overview
/introduction/.

171 **costs of vehicle air pollution:** Sarah E. Zelasky and Jonathan J. Buonocore, "The Social Cost of Health- and Climate-Related On-Road Vehicle Emissions in the Continental United States from 2008 to 2017," *Environmental Research Letters* 16, no. 6 (June 10, 2021), https://doi.org/10.1088/1748-9326/ac00e3.

174 **average American drives 13,476 miles a year:** US Department of Transportation Federal Highway Administration, "Average Annual Miles per Driver by Age Group," table, https://www.fhwa.dot.gov /ohim/onh00/bar8.htm.

174 **rural drivers tend to go farther:** Susan Meyer, "Average Miles Driven per Year in the U.S.," Zebra, August 31, 2023, https://www .thezebra.com/resources/driving/average-miles-driven-per-year.

174 **four separate trips a day:** "National Household Travel Survey Daily Travel Quick Facts," Bureau of Transportation Statistics, May 31, 2017, https://www.bts.gov/statistical-products/surveys /national-household-travel-survey-daily-travel-quick-facts.

175 **50 to 100 miles less than once a month:** Loren McDonald, "99.2% of US Daily Trips Are Less Than 100 Miles," Evstatistics, December 13, 2021, https://evstatistics.com/2021/12 /99-2-of-us-daily-trips-are-less-than-100-miles/.

CHAPTER 10: THE WALLET BALLOT

196 **Fashion waste has been accelerating:** *A New Textiles Economy: Redesigning Fashion's Future*, Ellen MacArthur Foundation, 2017, https://ellenmacarthurfoundation.org/a-new-textiles-economy.

196 **heat-trapping pollutants generated by fashion:** Valentina Portela, "The Fashion Industry Waste Is Drastically Contributing to Climate Change," California Public Interest Research Group, March 9, 2021, https://pirg.org/california/articles/the-fashion-industry-waste-is -drastically-contributing-to-climate-change/.

196 80 percent of Amazon rainforest deforestation: George Varagiannis, "How Leather Supply Chains around the Globe Are Tied to Deforestation," Collective Fashion Justice, August 14, 2022, https://www.collectivefashionjustice.org/articles/leather-lobbying-and-deforestation.

204 climate change—which they knew was coming: Neela Banerjee et al., *Exxon: The Road Not Taken*, Inside Climate News, 2015, https://insideclimatenews.org/book/exxon-the-road-not-taken/; Shannon Hail, "Exxon Knew about Climate Change Almost 40 Years Ago," *Scientific American*, October 26, 2015, https://www.scientificamerican.com/article/exxon-knew-about-climate-change-almost-40-years-ago/.

CHAPTER 11: SCHOOLED

208 called the Morris Model: "Morris Model Goals," Morris Model (website), https://www.morrismodel.org/morris-goals; "Strategic Plan," Morris Community Climate Smart Municipality Strategic Planning Retreat, October 29 and 30, 2018, https://www.morrismodel.org/_files/ugd/c432de_7953151f80bf4a43ba f46211a29248a8.pdf.

208 *Sierra* magazine's "Cool Schools" rankings: Katie O'Reilly, "The Top 20 Coolest Schools 2021," *Sierra*, September 9, 2021, https://www.sierraclub.org/sierra/cool-schools-2021/top-20-coolest-schools-2021.

209 twelve US colleges and universities have become carbon neutral: "Carbon Neutral Colleges and Universities," Second Nature Climate Leadership Network, https://secondnature.org/climate-action-guidance/carbon-neutral-colleges-and-universities/.

210 "Green Ribbon" school: Fryberger Elementary School, "School Nominee Presentation Form," US Department of Education Green Ribbon Schools, https://greenstrides.org/sites/default/files/webform

/CA_3_Disadvantaged_Fryberger_Elementary.pdf; "2021 U.S. Department of Education Green Ribbon Schools, District Sustainability, and Postsecondary Sustainability Award Honorees," U.S. Department of Education, https://www2.ed.gov/programs /green-ribbon-schools/2021-schools/awards.html.

215 **the "Montana 16":** Sam Bookman, "Held v. Montana: A Win for Young Climate Advocates and What It Means for Future Litigation," Environmental & Energy Law Program, Harvard University, August 30, 2023, https://eelp.law.harvard.edu/2023/08 /held-v-montana/.

215 **consider Nalleli Cobo:** "2022 Goldman Prize Winner Nalleli Cobo," Goldman Environmental Prize, https://www .goldmanprize.org/recipient/nalleli-cobo/#recipient-bio; Jen Wang, "Our Future: Billie Eilish on Climate Activism and Radical Hope," *Vogue*, January 4, 2023, https://www.vogue.com/article/billie -eilish-climate-activism-january-cover-2022-video.

Index

About the Author

Edward Humes is a Pulitzer Prize–winning journalist and author whose sixteen previous books include *The Forever Witness*, the PEN award–winning *No Matter How Loud I Shout*, and *Garbology*. He lives in Southern California with his human and greyhound and collie family.